THE MANAGEMENT OF QUALITY IN CONSTRUCTION

OTHER TITLES FROM E & FN SPON

European Construction Quality Standards
G. Atkinson

Profitable Practice Management
For the construction professional
P. Barrett

Construction Conflict Management and Resolution
P. Fenn and R. Gameson

The Foundations of Engineering Contracts
M. O'C. Horgan and F. R. Roulston

Project Control of Engineering Contracts
M. O'C. Horgan and F. R. Roulston

Competitive Tendering for Engineering Contracts
M. O'C. Horgan

Project Management
A systems approach to planning, scheduling and controlling
3rd edition
H. Kerzner

Managing Construction Worldwide (3 volume set)
Edited by P. Lansley and P. A. Harlow

Project Management Demystified
Today's tools and techniques
G. Reiss

Effective Speaking
Communicating in speech
C. Turk

Effective Writing
Improving scientific, technical and business communciation
2nd Edition
C. Turk and J. Kirkman

Journals

Construction Management and Economics
Editors: Ranko Bon and Will Hughes

Building Research and Information
(formerly Building Research and Practice)
Consulting Editor: Anthony Kirk

For more information on these and other titles please contact:
The Promotion Department, E & FN Spon, 2–6 Boundary Row, London, SE1 8HN.
Telephone 071-522 9966.

THE MANAGEMENT OF QUALITY IN CONSTRUCTION

J. L. ASHFORD

E & FN SPON
An Imprint of Chapman & Hall
London · Glasgow · New York · Tokyo · Melbourne · Madras

Published by E & FN Spon, an imprint of Chapman & Hall, 2-6 Boundary Row, London SE1 8HN

Chapman & Hall, 2-6 Boundary Row, London SE1 8HN, UK

Blackie Academic & Professional, Wester Cleddens Road, Bishopbriggs, Glasgow G64 2NZ, UK

Chapman & Hall, 29 West 35th Street, New York NY10001, USA

Chapman & Hall Japan, Thomson Publishing Japan, Hirakawacho Nemoto Building, 6F, 1-7-11 Hirakawa-cho, Tokyo 102, Japan

Chapman & Hall Australia, Thomas Nelson Australia, 102 Dodds Street, South Melbourne, Victoria 3205, Australia

Chapman & Hall India, R. Seshadri, 32 Second Main Road, CIT East, Madras 600 035, India

First edition 1989
Reprinted 1990, 1992

Typeset in 10/12 Garamond
Disc conversion by Columns Typesetters of Reading
Printed in Great Britain by TJ Press (Padstow) Ltd, Padstow, Cornwall

ISBN 0 419 14910 4

A catalogue record for this book is available from the British Library

Library in Congress Cataloging in Publication Data

Ashford, John L., 1930–
The management of quality in construction/
John L. Ashford
p. cm.
Bibliography: p.
Includes index.
ISBN 0 419 149104
1. Building — Quality control.
I. Title.
TH437.69 1989
690′.2 — dc19

We've gut to fix this thing for good an' all;
It's no use buildin' wut's a-goin' to fall.
I'm older'n you, an' I've seen things an' men,
An' my experunce, – tell ye wut it's ben:
Folks thet wurked thorough was the ones thet thriv,
But bad work follers ye ez long's ye live;
You can't git rid on't; jest ez sure ez sin,
It's ollers askin' to be done agin.

LOWELL, Biglow Papers

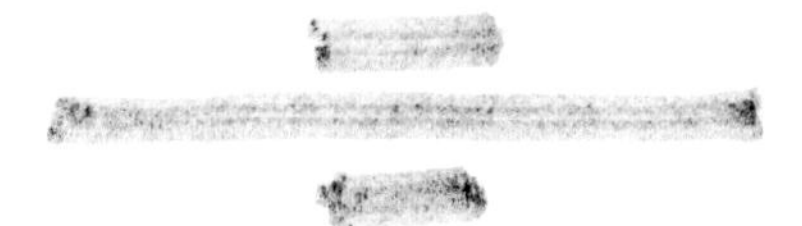

CONTENTS

ACKNOWLEDGEMENTS

The author wishes to thank his many colleagues in the Wimpey Group for their encouragement and assistance in the preparation of this book. Thanks are particularly due to Patricia Amoore and Lynne Quarrell for their efforts and patience in typing and re-typing the text. Extracts from British Standards are given by kind permission of the British Standards Institution, from whom official copies of the standards may be obtained. Their order department is at Linford Wood, Milton Keynes, MK14 6LE.

PREFACE

The purpose of this book is to provide a background of understanding of the principles of quality management for the benefit of those called upon to bear executive responsibility in the construction industry. It is not intended to be a specialist treatise — such would bring it into competition with the many expert and erudite works already in existence. More to the point, to treat the management of quality as if it were the province of specialists would run counter to the author's conviction that the principles and techniques which form his subject matter are indispensable tools for all managers.

Managers and engineers in the construction industry are busy people. They are men of action, although often traditional in outlook. They tend to have little time for theories or textbooks. So, what are the benefits they may expect to gain from the adoption of quality management?

For a start, they will be able to make a satisfactory response to clients who make the implementation of an effective quality system a condition of contract. This is not the best argument for adopting quality management, but it is probably the one which many companies find the most compelling in the first instance. The companies who benefit most from quality management, however, are those who do so for the purpose of improving their own efficiency by eliminating the costs, delays, waste, aggravation and disruption brought about by failure to do things properly first time. This is an aim worth pursuing for its own sake, irrespective of the demands of any particular client. If an organization can succeed in doing this, and at the same time can prepare itself to comply with mandated quality systems, then it will have achieved the best of both worlds. The intention of this book is to demonstrate that this can, and indeed should, be done.

In conclusion, the author apologizes to any female readers who may take exception to the general use of the male gender. This is purely for the sake of simplicity, and all references to he, his or him should be taken to include women as well as men.

1

THE QUALITY MANAGEMENT PHILOSOPHY

The concept of quality

In everyday usage, the word 'quality' usually carries connotations of excellence. When Shakespeare wrote Portia's speech on the quality of mercy, he described the particular characteristics of the subject which rendered it especially valuable. We speak of 'quality newspapers' when we wish to identify those which are known to provide material which will be of interest to an educated and discerning readership. In more class-conscious days, people referred to as 'the quality' were those held to be of high social status or of good breeding.

For reasons which will become evident, in this book, 'quality' will not be indicative of special merit, excellence or high status. It will be used solely in its engineering sense in which it conveys the concepts of compliance with a defined requirement, of value for money, of fitness for purpose, or customer satisfaction. With this definition, a palace or a bicycle shed may be of equal quality if both function as they should and both give their owners an equal feeling of having received their money's worth.

Quality, then, is a summation of all those characteristics which together make a product acceptable to the market. It follows that products which are lacking in quality will in the long term prove unmarketable, and that the purveyors of such products will go out of business. This truth applies not just to manufactured articles, it is equally valid when applied to services such as retailing, tourism or the practice of medicine. So the need to promote and control quality is of fundamental importance to any enterprise. Only by providing consistent value for money to their customers can companies hope to generate steady profits for their shareholders and ensure secure livelihoods for their employees. On a wider scale, the same observations can also be applied to nations. Those which have developed a reputation for quality products also have low rates of inflation, low unemployment, stable currencies and high rates of growth.

Paradoxically, the same market forces which in the long term permit the survival only of those who satisfy their customers can also tempt the dishonest or unwary to achieve short term profits by deceiving their customers with sub-standard products. Such practices are insidious and eventually fatal, and their prevention requires management action no less determined and formalized than is customarily applied to the control of money. These determined and formalized management actions are our subject matter.

The achievement of quality

With the meaning of quality which has been selected, it follows by definition that every purchaser of goods or services wishes to maximize the quality of his or her purchases. It also follows that the suppliers of such goods and services, if they wish to remain in business, must also ensure the quality of their products. Harsh experience of life, however, tells us that some suppliers put immediate profits ahead of long-term survival, and it is incumbent on every purchaser to protect himself from the dishonest or negligent supplier. In legal terms this is the principle of *'caveat emptor'*, or 'let the buyer beware'.

But few organizations are solely either buyers or sellers. Those who sell also buy and those who buy also sell. In the construction industry the client or purchaser of a project has to sell its benefits to his customers, be they tenants, taxpayers or other kinds of consumer. The contractor who sells a project to his client is a buyer of materials, labour and sub-contracts. The consulting engineer or architect who sells design has to buy the services of professional designers and draftsmen. As long as each buyer in the chain is able to obtain an acceptable level of value for money, and as long as the sellers are able to provide that level and still make a profit then the market is able to function.

Chains of buyers and sellers are also to be found within organizations. For example, in a housebuilding company the marketing department supplies information on market requirements to the architects. The architects' department 'sells' drawings and specifications to the construction department. The construction department supplies completed houses to the sales negotiators. Each group relies on one or more internal 'suppliers' to provide the information or materials it needs so that it can then in its turn satisfy the requirements of its customers, both internal or external.

What means are available to enable a purchaser to assure the quality of a purchase? In early village economies, it was comparatively easy. The farmer wishing to buy a cart would approach a local cartwright and agree a specification and price. The cartwright would apply traditional skills in design and in the selection of materials. The workmanship would be that of himself or of apprentices under his personal control. Pride of craftsmanship and the need to preserve a reputation in the locality would effectively prevent careless or deliberate flouting of accepted standards. The coming of the Industrial Revolution in the early nineteenth century rendered this relatively simple quality system obsolete. With the development of new sources of power, it became more economic to concentrate production into factories and distribute the products of the factories to the purchasers by canal or railway. Purchasers became separated from suppliers and the imperatives of craftsmanship and the preservation of local reputation lost their power.

Early factories were dirty, dangerous and hopelessly inefficient, but the demands for their products increased rapidly. This was particularly so in America, where swelling consumer demands for manufactured goods coincided with an unending influx of desperately poor, but unskilled, immigrants arriving from Europe and seeking work in the factories. In the 1880s, Frederick W. Taylor (1856–1915) studied the organization of work in factories. His purpose was to make the manual worker more productive, and therefore better paid, and at the same time to relieve him of unnecessary and wasteful labour. His method was to identify and analyse all the operations which had to be performed for a given task and then to optimize the sequence of operations to create the smoothest and most economical flow of work. The consequent development of the techniques of mass production has probably contributed more than any other factor to the increase in the affluence of ordinary people in the last century.

Taylor observed that those who knew what was to be done seldom knew how it should be done. He recognized that planning and doing were separate activities, that the one should precede the other, and that planning would not happen if mixed in with doing. On the other hand if planning were allowed to become entirely divorced from doing, it would cease to be effective and this could become a threat to performance. Unfortunately in all too many cases, this is exactly what happened. So, with purchasers separated from producers and planners separated from doers, what happened to quality? Obviously, it suffered. At first this did not matter very much. The products being made were simple and uncomplicated, and the consumers were easily satisfied – Henry Ford offered them any colour of car they liked, as long as it was black. Nevertheless, it was accepted that some control of quality was necessary and inspectors were appointed for this purpose.

In the United Kingdom, one of the first large scale applications of the Taylor principles was in armaments manufacture during the First World War. A largely female work force, quite unskilled and without any previous experience, was drafted to the ordnance factories. They were taught simple repetitive tasks by the few male craftsmen who remained, and set to work. The craftsmen became inspectors, checking what was produced and rejecting products which were not up to specification. The efforts of the workers were heroic, and the remarkable outputs they achieved were a vital element in the war effort, but the system was neither efficient nor cost effective. Lying at the heart of this inefficiency was the belief that quality could be adequately controlled solely by inspection. This, unfortunately, was not the case.

The inspection of quality

The effect of the Taylor system was to replace the traditional skills of the self-employed and self-motivated craftsman with those of the production engineer. It achieved spectacular increases in productivity, but at a price. Firstly, by separating planning from execution, it deprived workers of the right to decide how their work should be done and at what rate. Secondly, the organization of work into highly repetitive short-cycle activities made the lives of the workers boring, monotonous and devoid of meaning. Inevitably quality suffered.

The response of management was to establish independent internal inspection departments to restore the balance. Inevitably, since they had to be capable of distinguishing between what was satisfactory and what was not, the inspectors were drawn from the more intelligent and knowledgeable members of the work force. This had the immediate effect of diluting the skills of the productive team and at the same time creating a powerful group whose existence could be justified only if it found work which had to be rejected. It is difficult to imagine any arrangement more calculated to de-motivate the productive work force and create friction and bad feeling.

The existence of inspection departments removed any remaining feeling of responsibility on the part of the work force for the quality of their work. The criterion of acceptability changed from 'Will it satisfy the customer?' to 'Will it be passed by the inspector?'. Ways were found to deceive or bypass the inspectors, who retaliated by requiring more frequent and more stringent testing. There had to be enough inspectors to accommodate peaks in workload so as to avoid the inspection department being accused of holding up production. At off-peak periods the inspectors became under-employed and occupied their time by creating bureaucratic procedures to further strengthen their stranglehold on their enemies in the production departments.

An inspector can identify a fault only after it has been committed. He may then order the item to be scrapped or rectified in some way. Whatever the decision, waste will have occurred and harm done which cannot be undone. In many cases the inspector will know the cause of the fault and how it can be prevented, but he has no incentive to pass on this knowledge to those in charge of production. They are on the opposing side and are unlikely to welcome his advice.

Such then, are the problems of attempting to control quality by inspection alone. There has to be a better way, and there is.

The management of quality

After the Second World War the economy of Japan was in ruins. To attain their military objectives, all available resources of capital and of technical manpower had been directed to armaments manufacture, while their civilian economy gained an unenviable reputation for producing poor-quality copies of products designed and developed elsewhere. Unless they were able to raise the quality of their products to a level which could compete, and win, in the international market place they stood no chance of becoming a modern industrialized nation.

To learn how to regenerate their industries, they sent teams abroad to study the management practices of other countries and they invited foreign experts to provide advice. Among the latter were two Americans, J.M. Juran and W.E. Deming, who brought a new message which can be summarized as follows:

1. The management of quality is crucial to company survival and merits the personal attention and commitment of top management.
2. The primary responsibility for quality must lie with those doing the work. Control by inspection is of limited value.
3. To enable production departments to accept responsibility for quality, management must establish systems for the control and verification of work, and must educate and indoctrinate the work force in their application.
4. The costs of education and training for quality, and any other costs which might be incurred, will be repaid many times over by greater output, less waste, a better quality product and higher profits.

These are the basic principles of the management concepts which have since become identified under the generic term of quality management.

The Japanese developed and refined what they had learnt, adapting it to their own environment and to the circumstances of individual companies. They made the management of quality an integral part of the manufacturing process and proved that by reducing the incidence of defective goods, the costs of production can be decreased substantially. Armed with the techniques of quality management, the Japanese proceeded to achieve virtual world domination in a series of key industries. At the time, who would have guessed that thirty years later and with their economy under siege, the Americans would be desperately trying to re-learn the lessons of quality management from their erstwhile pupils?

Meanwhile, back in the United Kingdom, the Ministry of Defence was facing increasing difficulties in the procurement of military equipment. Manufacturers still operated on Taylor principles and despite internal inspection, failures of equipment when in service were at an unacceptably high level. For many years the Ministry had also put its own inspectors into the factories and used the services of third-party inspection bodies, but this did not solve the problem. The presence of external inspectors served to remove the sense of responsibility of the internal inspectors, just as the internal inspectors had removed responsibility from those doing the work. Armaments were becoming more and more complex and in spite of all the inspectors, failures in service were frequent, expensive and dangerous.

In May 1968, the North Atlantic Treaty Organization (NATO) issued the first edition of an Allied Quality Assurance Publication known as AQAP-1. Entitled *'Quality Control System Requirements for Industry'*, this document specified NATO requirements for quality control systems to be operated by their contractors. These systems were required to serve two purposes: firstly to ensure that goods and services conformed to contract requirements and, secondly, to provide objective evidence of such conformance. When invoked in a contract or purchase order, AQAP-1 had mandatory effect. To assist contractors and manufacturers to comply with its provisions and to guide those charged with evaluating quality systems, NATO issued a further document known as AQAP-2 in September 1968.

In response to NATO requirements the British Ministry of Defence published its own equivalents of the AQAP documents for use in the United Kingdom. The first of these documents was *Defence Standard (DEF STAN) 05-08*, issued in March 1970, but a more definitive document, *DEF STAN 05-21* was published in January 1973. Unlike many subsequent standards, *DEF STAN 05-21* was a model of clarity. The concepts upon which it was based are worth quoting:

'(1) The quality of manufactured products depends upon the manufacturer's control over his design, manufacture, and inspection operations. Unless a product is properly designed and manufactured it will not meet the requirements of the buyer. Accordingly manufacturers must be prepared to institute such control of quality as is necessary to ensure that their products conform to the purchaser's quality requirements.

(2) Manufacturers should be prepared, not only to deliver products on schedule at an agreed price, but in addition, to substantiate by objective evidence, that they have maintained control over the design, development, and manufacturing operations and have performed inspection which demonstrate the acceptability of the products. The design phase is considered to embrace all activities after the statement of the operational requirement, through to the point at which the requirement has been satisfied.'

These concepts, which incorporated some, but not all, of the principles expounded by Juran and Deming, signified an evolution of the principle of *caveat emptor*. With the increasing complexity and multiplicity of industrial processes, it was no longer possible to judge the acceptability or otherwise of a product by inspecting it in its finished state or even by a series of stage inspections. Instead, the assurance of compliance with specification that a purchaser legitimately demanded should be achieved by the appraisal, approval and surveillance of the supplier's management arrangements combined with spot checks and audits to prove that they were being implemented as agreed and were in fact achieving their objectives. To quote again from *DEF STAN 05-21*:

> 'For his protection the purchaser should exercise such surveillance over the manufacturer's controls, including inspection, as is necessary to assure himself that the manufacturer has achieved the required quality. Such surveillance should extend to sub-contractors when appropriate. The amount of surveillance performed by the purchaser is a function of the demonstrated effectiveness of the manufacturer's controls and of the demonstrated quality and reliability of his products. In the event that the purchaser's surveillance demonstrates that the manufacturer has not exercised adequate control the purchaser will have valid reason because of his contract stipulations to discontinue the acceptance of the product concerned pending action by the manufacturer to correct whatever deficiencies exist in his quality control system.'

The introduction of concepts of quality management into defence contracts prompted the British Standards Institution to take action to provide guidance and information on the subject to a wider industrial audience. In 1971 they published BS 4778: *Glossary of terms used in quality assurance* and followed this in 1972 with BS 4891: *A guide to quality assurance*. Although they were only advisory, these documents served a useful purpose in interpreting the requirements of defence procurement standards in a more general context. The process was completed in 1979 with the publication of the first version of BS 5750: *Quality Systems*. This served as the definitive standard in the United Kingdom until 1987 and was used as the basis of the International Organization for Standardization's ISO 9000 series of standards. BS 5750 was re-issued in 1987 in a form identical with the corresponding ISO standards.

These standards introduced the words 'quality system' into the language of management. They established that a quality system has to achieve two objectives – first it has to control what is produced to make sure it meets the requirements of the purchaser and, secondly, it has to provide confidence or assurance that compliance has been achieved. This confidence or assurance is needed by both the buyer and the seller, the former so that he knows he is getting what he is paying for and the latter so that he knows his system is working.

Standards and specifications for quality systems are important and their contents are discussed in detail in Chapter 3. Many companies and individuals make their first acquaintance with the subject of quality management when obliged to provide evidence of compliance with a quality system standard before or while tendering for a contract. In such cases, the standards are used to define actions to be imposed by one party on another. It is interesting to contrast this approach to the management of quality with that preached by Juran and Deming and practised to such good effect by the Japanese. The philosophy they propounded required that management should devote its attention to the improvement and maintenance of quality not because someone else might oblige them to do so, but because it was a desirable end in itself.

There is a significant difference between an organization which truly believes in the need to manage quality and one which merely prepares itself to comply with a standard. The latter will have the systems, procedures, manuals and so on which are required by the standard, but unless the people who have the task of operating the systems have the right attitudes and inner motivation, the results will not be wholly successful. These attitudes and motivation can be inculcated only by a long-term programme of company-wide quality improvement, initiated and supported by the overt personal involvement of the chief executive and his senior colleagues. The strategy and tactics of such a programme are the subject of Chapter 11.

Applications in construction

The early development work on quality management took place in a manufacturing environment and so it is hardly surprising that most literature on the subject is written in the vernacular of the factory. This is unfortunate as it creates a mistaken impression in the minds of those engaged in activities other than manufacturing that the tenets of quality management hold no benefits for them. Nothing could be further from the truth. Only monopolies can afford to ignore the customer's demand for value for money and satisfaction of his needs. Any person or organization whose livelihood depends on successful performance in the market place can benefit from quality management, and this includes the construction industry.

This is not to say that quality management as practised in factories can be transplanted unchanged into the construction industry. The differences between the factory and the construction site cannot be ignored. There are special factors which have to be taken into account — the susceptibility to weather, the mobility of labour, the fact that almost every job is a prototype, and so on. These realities undoubtedly make the introduction of quality management more difficult than in other industries. But if it is true that the management of construction sites is a uniquely formidable task, it does not make sense to ignore the most significant advance in management technique to have arrived on the scene in recent years. In any case, times are changing. The differences between factories and construction sites are slowly but steadily disappearing. There is more and more assembly work and less and less craftsmanship.

So the techniques are there, available for the industry to use. Are they needed? The answer has to be that they are. All too many buildings and structures in recent years have failed to satisfy the legitimate requirements of their purchasers and the reasonable expectations of the community at large. The record is not one of which any manager or engineer can feel proud. There are faults in concept, design, materials and workmanship. Some of these are a result of technical advances; structures which were once designed by experienced engineers and erected by highly trained craftsmen are now increasingly being designed by computers and assembled by semi-skilled labour. Analysis shows, however, that only a minority of construction defects are technical in origin. Far more arise from inadequacies in the management structure of the industry, from lack of training and from the commercial pressures which stem from the almost universal custom of awarding work only to the lowest bidder.

Recent advances in the techniques of construction management have increased the commercial pressures. More and more organizations are operating sophisticated computer-based programs for analysing and presenting the facts, statistics and trends upon which management decisions may be based. No longer need the manager, or his superiors, be unaware of potential delays or budget overruns or cash-flow problems. The computer will present all the information needed, and more, with speed, accuracy and detail. These systems are powerful motivators. They enable organizations to set their managers exact goals and then report on the achievements of these goals quickly and precisely. Managers soon come to realize that the way to achieve recognition, salary increases, promotion and all the other rewards of work, is to make certain that the information fed to the computer will be that which will tell their superiors what they want to hear. It would be naive not to recognize that the quality of the product is likely to be the first casualty as managers succumb to the pressures and priorities exerted by the system. If this is the case, control of total cost, which is the express purpose of most project management systems, will be frustrated. The products of the construction industry are intended for a long life and those who construct them can be held liable for defects for many years after hand-over. The normal time cycle of monetary control is too short to collect all the costs which will eventually accrue and which should rightfully be charged against the project or work element. Project management systems which present their targets only in terms of cost and schedule can destroy a manager's natural instinct to produce work of which he can be proud. This is a very strong argument in favour of the establishment of a quality system. It can provide a counterpoise to the two forces of cost and schedule which, if not resisted, will pull the site manager in the wrong direction — the analogy of the triangle of forces will be well understood by engineers accustomed to applying the laws of statics.

No one with experience of managing a construction site will underestimate the magnitude of the task. But, when one reviews the frequency of press reports of defective structures, when one considers the poor reputation of the construction industry for the quality of its products and, above all, when one contemplates the spiralling costs of litigation over latent defects, it can only be concluded that we do have a serious problem. The arguments in favour of applying the logic and rationality of quality management to the particular problems of the construction industry are very powerful. The following chapters will explain how this may be accomplished.

2

THE CONSTRUCTION TRADITION

So far, the subject of quality management has been discussed in a general industrial context. Before focusing on the specific requirements of the construction industry, it is necessary to digress briefly on the evolution of the industry to its present state and to look at traditional and existing methods of assuring quality.

The products of construction are expensive, complex, immovable and long-lived. They seldom offer scope for repetition, they have to be built where they are needed and, if not designed or built correctly, there is usually little that can be done to put things right at a later stage. Furthermore, it is only occasionally that the potential purchaser can examine the finished product before he has to commit himself to purchase. To overcome these difficulties, systems of contract have evolved to facilitate the formal commitments which are necessary before work can proceed. These contracts define what is to be built, the roles of the various parties concerned and the terms of the bargain between them. In so doing, they provide the framework of quality systems, perhaps not so detailed as those described in BS 5750 or DEF STAN 05-21, but quite effective none the less. They specify the purchaser's requirements, they stipulate the measures to be taken to assure compliance and they state the remedies available to each party in the event of default.

The contractual systems used in the construction industry reflect the fact that it is the inheritor of two quite separate and distinct traditions, one as old as civilization itself and the other dating back no further than the Industrial Revolution. These two traditions are that of the architect and builder, and that of the civil engineer. Most major construction contractors undertake both building and civil engineering work and there is considerable interchange of staff between the two disciplines. Likewise many of the larger consultancies include architecture, structural engineering and civil engineering among the services they offer. However, the contractual arrangements used for each discipline and the approaches they adopt towards the management of quality still reflect the different historical backgrounds.

Building and architecture

The purpose of a building is to provide shelter. The purpose of architecture is to inspire. Sometimes the two come together, and we have a building which not only provides comfort, convenience and security for its occupants, but also excites aesthetic sensations in those who enter or behold it.

Early buildings had few aesthetic pretensions. Stone Age customer specifications were no doubt simple: 'Buildings shall be warm, dry and capable of preventing entry by marauders and wild animals'. However, the more advanced civilizations eventually developed to a point at which they could sustain groups of people who had the time and leisure to look beyond their immediate survival needs and seek more spiritual forms of satisfaction. The Aztecs built temples, the Egyptians built pyramids, and the art of architecture was born. The Greeks, too, built temples. These were of great beauty, designed to comply with the assumed requirements of the gods to whom they were dedicated. From the outside, they were marvels of proportion, symmetry and ornamentation, but they were not built to be occupied by humans, indeed their doorways were often concealed by colonnades as if to discourage entry. Nevertheless, any functional limitations they may have had were outweighed by their aesthetic quality and they established a classical tradition of architecture which was to last for centuries.

It fell to the Romans to introduce the more mundane concepts of utility and customer satisfaction. They developed the principle of the arch to overcome the limits to column spacing imposed by the practical dimensions of a horizontal architrave, thus achieving new standards of internal space and access. The arch, and the vault which was developed from it, are engineering concepts, applicable not only to basilicas, bath-houses and theatres, but also to bridges, aqueducts and military fortifications.

The building of an arch or vault required stones cut precisely to shape and size. This was the craft of the stonemason and it demanded an understanding of three-dimensional geometry and a feel for the fracture characteristics of rock masses. The designers of buildings were initially drawn from the ranks of the master-masons. They carried their design knowledge in their heads, and it grew as they moved from one construction site to another. They were an intellectual elite, and the skills of their hands and brains were secrets to be closely guarded from outsiders. The development of the Roman arch to the Gothic arch, the flying buttress, the fan vault, and all the other architectural delights of the European ecclesiastical tradition were among the achievements of the master-masons. They became expert not just in their original craft of cutting and placing stone. They had also to design timber falseworks to support their stonework in its unfinished state, they could sculpt figures and carve ornaments, they devised machines for lifting heavy weights to great heights, and they hired, trained and managed the work force. They were the forerunners of today's design and construct contractors.

Few would dispute the quality of the surviving works of the great master-masons. No doubt they had their failures as well as their successes, but these we no longer see as they have either fallen down or have been pulled down. But who were the customers whose satisfaction would provide a measure of quality? The market of the master-mason was the patron. Many patrons were royal, others were ecclesiastical, political or commercial. They required palaces, cathedrals, town halls and mansions. The purpose of these buildings was not just to provide living accommodation or meeting places, but also to symbolize the power, spirituality or wealth of the patron. Their quality had to be judged not in terms of durability, serviceability or economy, but on their ability to impress and inspire. The latter is the prerogative of the artist, and the master-masons most in demand were those who were able to combine technical skills with artistry, that is, to be architects.

It is a remarkable feature of early buildings that their designers obeyed basic laws of mechanics, the principles of which were not scientifically formulated until long after the buildings were constructed. For example, it is difficult to imagine how one could successfully design and construct a twelfth-century Gothic cathedral without a clear understanding of the nature of gravitational force and the laws of statics. Yet these matters were not fully rationalized and codified until the works of Galileo and Newton were published some five hundred years later. The great cathedrals were exceptional in their technical demands. Most buildings were more domestic in scale and conventional in construction, posing few structural problems which could not be solved using empirical methods. Architecture as an art form flourished during the Renaissance period — indeed it is remarkable how many of the famous architects of the time had established reputations in other artistic fields before becoming architects.

However, the concentration of power into the hands of national governments in the early seventeenth century brought about a change. Art and architecture became integral parts of the system of government.

Architecture was useful to create work and to celebrate the greatness of king and state. The French, as so often to the forefront in the pursuit of national prestige, led the way and established an academy for architecture. Those who achieved distinction therein became royal architects, well paid members of the court, but dependent on the patronage of the state. Thus was established the formal qualification of the architect, no longer a self-made amateur, but with his status and authority protected by statute. This development signalled the separation of the design of buildings from their construction. The architect became a design expert, but he no longer had mastery of the techniques of construction — these remained the preserve of the stonemason, the carpenter and the plumber. The architect was trained to establish the shape of a building, the allocation of space, the surface finishes and textures, the ornamentation and so on, but the builder was expected to attend to the details of waterproofing, durability, economy and serviceability, and to control the construction site.

Until the nineteenth century, the predominant construction material for buildings was stone. Masonry structures can resist only compressive forces, and even the most elaborate require little more than an understanding of the triangle of forces for their analysis. However, the introduction of wrought iron, and after 1860 of steel, created new problems. These materials can resist tension, indeed this is their purpose. The analysis of structures incorporating members both in compression and tension cannot satisfactorily be undertaken by empirical methods, but requires the special skills of the structural engineer. Thus began a trend towards increasing specialization in building design. Most architects failed to come to terms with the new technologies and as a result they ceased to be the sole authorities on design matters. The designers of major industrial structures such as textile mills and railway stations, were engineers, not architects. Sir Joseph Paxton, who designed the Crystal Palace, one of the most significant architectural innovations of the nineteenth century, was a distinguished gardener and horticulturist who had become skilled in the use of iron and glass in building conservatories and greenhouses.

In modern times we have seen the development of the skyscraper, factory-made buildings, new plastic materials and air-conditioning. No longer are we obliged to design our structures from the materials available, we can design the materials for our structures. The design and specification of buildings has come to require a synthesis of architecture and engineering. The architect still leads the design team, determining the concept, the proportions and style of the building, but he increasingly relies upon the technical skills of the structural engineer, the materials engineer, the building services engineer, and last but by no means least, the practical know-how of the builder.

Civil engineering

Thomas Telford was born in 1757. On leaving school he was apprenticed to a local stonemason. An ambitious youth, he left his native Eskdale after completing his apprenticeship and sought work, at first in Edinburgh and a year later in London. His ambition was to be an architect.

On arrival in London, Telford was fortunate to be introduced to Sir William Chambers, who at that time shared with Robert Adam the Royal office of 'Architect of the Works', and was later to become a founder of the Royal Academy. Telford was set to work by Chambers on the construction of the new Somerset House, trimming and setting blocks of Portland stone. Although the excellence of his work led to promotion, this was inadequate reward for the ambitious Telford. He sought commissions for house improvements and was successful in obtaining work on Westerhall House in Eskdale and the vicarage of Sudborough in Northamptonshire. These were minor works, but they led to another and far more important commission. This was to superintend the erection of a new Commissioner's house, a chapel and other buildings in Portsmouth Dockyard. It was at this time in 1786 that Telford is reported to have commenced the programme of scientific study which led to his later eminence.

In Britain in the late eighteenth century there were no schools of architecture to match those established in France in the seventeenth century. It was not until 1841 that the University of London established the first chair of architecture in a British university. The function of an architect was undefined, and architects, contractors and craftsmen all undertook such tasks as they (or their customers) considered themselves capable. The architect for the work at Portsmouth Dockyard was one Samuel Wyatt, brother of the more famous architect James Wyatt. Samuel Wyatt had held the carpentry contract at Somerset House, and may have been influenced in arranging Telford's appointment for the work at Portsmouth. Telford had, not long before, contemplated setting up in business as a building contractor but had to abandon the idea because of lack of working capital. At Portsmouth he is reported to have 'superintended construction' and was presumably responsible for the quality of materials and workmanship, but he does not appear to have had money of his own at risk.

If the role of the professional architect in the Britain of 1786 was confused and ill-defined, that of the engineer was hardly recognized at all. This is not to say that major engineering works had not been undertaken — the canals of James Brindley and others were proof of this. But James Brindley was a millwright by trade and the term 'engineer' was seldom used except in the military context. The builders of canals, and later of railways, were brash, self-made men, lacking the social graces and theoretical knowledge of their continental counterparts, but making up for this with practical trade experience and the ability to hire and control large teams of men. They were not held in high regard by society at large and the profession of architect, which did at least carry the cachet of being an applied art, was the target of those in the construction industry who wished to rise in the world.

Telford's next post was as Architect and Surveyor of Public Works for the County of Salop. He was then aged 30. His responsibilities were wide ranging: he built a new infirmary and a county gaol in Shrewsbury, he designed and built churches, he undertook what we now call town-planning, and he built bridges.

Major bridges in 1787 were built of stone, by stonemasons. They consisted of arches or vaults, were surmounted by parapets or balustrades and were decorated with many of the architectural features normally found on buildings. Thus it is not surprising that bridge design should have been considered the province of the architect. The first cast-iron bridge, at Coalbrookdale, was designed by an architect, Thomas Pritchard. While of great historical interest, this was a clumsy design, imitative of timber structures and its producer failed to take advantage of the particular properties of the new medium. When Telford was offered a similar opportunity to design a cast-iron bridge at Buildwas, just upstream of Coalbrookdale, he responded with a superior design, having an arch lighter, flatter and infinitely more graceful than that produced by Thomas Pritchard. He achieved this success because he was able to apply scientific theory to a novel and untried material and thereby create a new structural concept. Telford had become the first modern British civil engineer.

The term 'civil engineer' in Telford's time was all-embracing and meant a civilian, in contrast to a military, engineer. With the progress of mechanical and electrical engineering during the nineteenth century these came to be regarded as separate branches of engineering, and civil engineering eventually acquired the more restricted meaning with which it is now associated. Telford was self-taught: engineering first became a subject for tuition in England at Cambridge in about 1796, some fifty years after the foundation of the Ecole des Ponts et Chaussées in Paris, and it was not until 1838 that King's College, London established the first chair in civil engineering. Despite his lack of formal education, Telford went on to build many more bridges, roads, canals and a huge variety of other works. But the physical structures he built were not his only achievement. He also established procedures for the execution of civil engineering projects which still remain the basis of current contractual arrangements.

Construction contracts

This book is not written by a lawyer and it is not the intention to provide a treatise on contract law as it is applied in the construction industry. However, the function of the contract in the management of quality is of such importance that it would not be possible to deal adequately with the latter without ensuring that the reader has a sufficient understanding of the former. The paragraphs which follow give a general description of conventional contractual arrangements, but it should be noted that there is a wide range of alternatives which may be found more suitable for particular applications.

The essence of any contract is that two parties together make a bargain whereby one party promises to provide some consideration or payment in exchange for some thing or service offered by the other party. Contracts may be in many forms and do not have to be documented to be enforceable, but, in the case of construction works, there is usually so much money at stake that a written contract is almost always considered desirable by both parties.

The party to a construction contract who makes the payment is the client, who may also be referred to as the employer, the owner or the purchaser. In most quality system standards he is known as the purchaser, so, to be consistent, this term will be used hereafter. The purchaser may be a private individual, a limited liability company, a local authority, a government department or any other incorporated or unincorporated body. The other party to the contract, who is to carry out the works, is the contractor, otherwise known as the builder, building contractor or civil engineering contractor. The purchaser and the contractor are the two parties to the main contract, which usually follows one of the standard forms of contract in common use within the industry.

Clearly there has to be a mutual understanding between the two parties to the main contract as to what is to be built in return for the agreed consideration. This requires that the contract should incorporate a set of drawings and a specification. For a purchaser having architectural or engineering expertise within his own organization, this is no problem — his people will prepare the necessary documents and seek tenders from interested contractors. However, purchasers who only rarely enter the market for construction projects are not likely to have these resources at their disposal and they will find it necessary to enter into a separate contract with an architect or a consulting engineer, or perhaps both, to carry out the design work and to assist in supervising the execution of the main contract. Again, there are standard forms of contract for the engagement of architects and engineers and these stipulate the role and responsibilities of the person

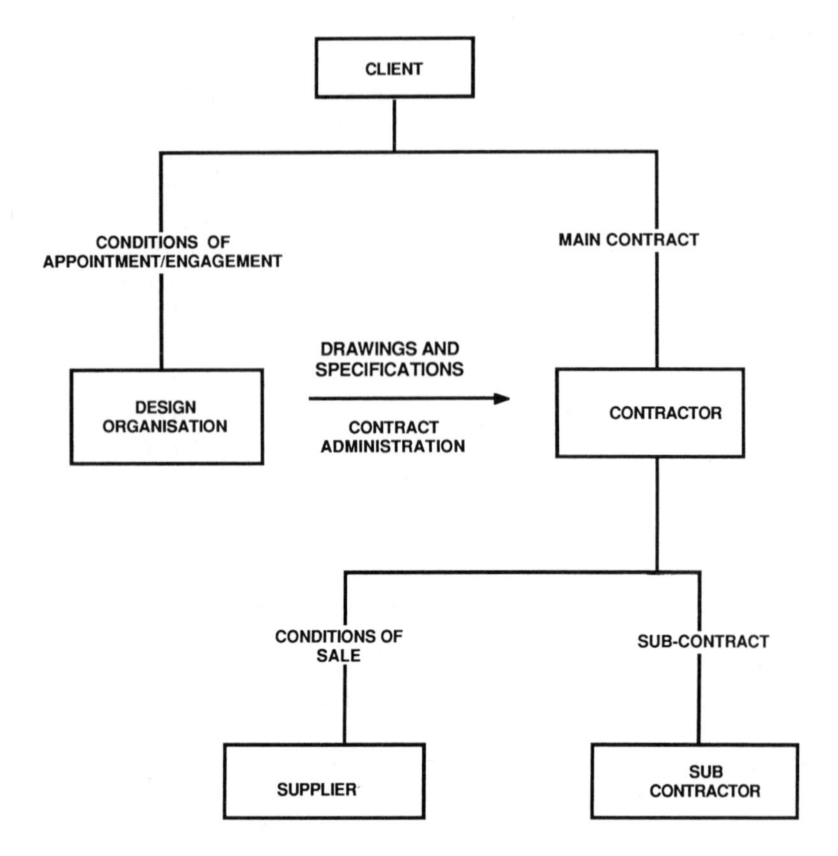

Figure 2.1 Conventional construction contracts.

or persons engaged and the fees which will be paid. These conventional contractual arrangements are illustrated diagrammatically in Figure 2.1.

During the design stage the consulting engineer or architect acts as a contractor to the purchaser, but once it is proposed to construct the works he may also become an agent, acting on the purchaser's behalf to supervise and administer the work of the construction contractor. His contract with the purchaser normally requires that he should act in a fair and impartial manner in his dealings with the two parties to the main contract. However, as an agent of the purchaser, he is also required, if only implicitly, to act in the purchaser's interests. A staff architect or engineer in the employ of the purchaser is expected to observe a similar duality of responsibility. This is a remarkable arrangement, seldom encountered outside the construction industry, and the fact that it can be sustained, even when the individual concerned is an employee of the purchaser, reflects great credit on the integrity of the professionals concerned.

Returning to the main contract, that between the purchaser and the construction contractor, the standard conditions of contract most frequently encountered are known respectively as the JCT Standard Form of Building Contract and the ICE Conditions of Contract. The JCT form of contract, which is intended for use with all types of building work, is issued by the Joint Contracts Tribunal, whose constituents include the Royal Institute of British Architects (RIBA), the Royal Institute of Chartered Surveyors (RICS), the Building Employers Confederation (BEC) and other bodies representing local authorities and contractors. The ICE form of contract, sponsored jointly by the Institution of Civil Engineers, the Federation of Civil Engineering Contractors and the Association of Consulting Engineers is designed for use in connection with civil engineering works.

Before considering the differences between these two forms of contract, let us look at aspects in which they express a common approach to matters which affect the management of quality. Although perhaps worded in different ways, both forms of contract provide for the following:

1. The nomination of the person, or sometimes the firm, who will represent the purchaser in his relationship with the contractor. This may be an employee of the purchaser, or it may be an architect or consulting engineer engaged to design and supervise construction of the works. It is a convention that the purchaser's representative is identified in contract documentation by heading his title with a capital letter. In the JCT form he is known as the Architect, if the person concerned is registered as such under the Architects Registration Act 1938 or, if not, as the Supervising Officer (SO). Under the ICE Conditions of Contract the appointee is known as the Engineer. The term 'engineer' is not protected by statute in the United Kingdom.
2. The appointment by the contractor of an agent to be in charge of the works and to receive instructions given by the Engineer or the Architect/SO on behalf of the purchaser.
3. The stipulation that the contractor must complete the works by a certain date, in accordance with the contract drawings and specifications and to the satisfaction of the Architect/SO or the Engineer.
4. The establishment of the right and duty of the Engineer or the Architect/SO to have access to the site and to inspect and examine the progress and quality of the works.
5. The duty of the Engineer or the Architect/SO to certify payments to the contractor for work in progress and for materials delivered to the site.
6. The prohibition of the contractor from sub-contracting sections of the work to others without the consent of the Engineer or the Architect/SO.
7. The provision of a maintenance period (normally one year) after completion of the works during which the contractor is required to make good any defects. To make sure that the contractor honours this

obligation, a retention may be subtracted from interim payments and held back until the contractor has satisfactorily discharged his responsibilities for maintenance.

8. The settlement of disputes by arbitration.

So much for the similarities between standard forms of contract for civil engineering and building works. There are also substantial differences, and these centre largely upon the respective powers of the Engineer and the Architect. In general terms, the Architect has fewer powers and responsibilities than does the Engineer. This is partly a matter of tradition, but it is also a reflection of the fact that the decisions of an architect rarely concern matters which might expose the general public to physical hazard, whereas the converse applies to the decisions of an engineer.

The following paragraphs are based on the requirements of the two standard forms of contract used for the engagement of architects and engineers respectively. These are the Conditions of Appointment recommended by the Royal Institute of British Architects (RIBA) and the Conditions of Engagement published by the Association of Consulting Engineers (ACE).

Responsibilities of the Architect

An Architect who accepts a commission under the RIBA Conditions of Appointment will first produce an outline design for the scheme or project under consideration. He will seek the purchaser's approval of his proposals and will then develop the design, making sure that it complies with all the relevant Building Regulations and other statutory requirements. He will prepare the tender documents and send out invitations to tender. On receipt of tenders, he will evaluate them and advise the purchaser on which should be accepted. He will then provide such additional information as the contractor may reasonably require, visiting the site as appropriate to monitor the progress and quality of work. He will certify payment for completed work and keep the purchaser informed on the financial status of the project. On completion, he will give general guidance on maintenance and provide the purchaser with a set of as-built drawings.

This is an extensive list of duties, but the Architect does not act in isolation. Indeed, one of his first responsibilities under the contract is to advise the purchaser on the need for other consultancy services and the scope of these services. These consultants may include:

1. A quantity surveyor to prepare bills of quantities, to measure work done and to prepare valuations for certification.

2. A structural engineer to design the foundations and structural framework of the building and to supervise their construction.
3. An engineering systems (or building services) engineer to design and supervise the installation of equipment such as boiler plants, heating and ventilating equipment, electrical distribution systems, lifts, etc.

If the purchaser agrees to the need for these specialists, he and the Architect will identify mutually acceptable individuals or firms who will then be appointed, either directly by the purchaser or through the agency of the Architect. Thereafter the Architect will be responsible for co-ordinating the work of the specialists and integrating the contributions of each designer into the overall concept. However, and this is most important, each consultant is responsible to the purchaser for the quality of his work, not to the Architect. Thus the Architect will not normally be held liable under the contract for negligence on the part of other members of the professional team.

During construction, the Architect will visit the site to determine that the works are being executed generally in accordance with the contract drawings and specifications. If the Architect and the purchaser are of the opinion that more frequent or constant inspection is needed then an inspector or 'clerk of works' will be employed. The clerk of works will normally be employed by the purchaser, but he will operate under the direction and control of the Architect. The basic duty of a clerk of works is to be present on the site, to watch and record what is done and to report the facts to the purchaser and to the Architect. Directions given to the contractor by a clerk of works are valid only if related to a matter on which the Architect is contractually empowered to give directions and if confirmed in writing by the Architect within two working days. It says much for the personal qualities and experience of many clerks of works that, despite these rather limited powers, they frequently earn respect and influence on the site far in excess of that provided for in the contract.

So, what are the Architect's contractual responsibilities for the quality of construction? The RIBA Conditions of Appointment provide for periodic, but not constant, supervision by the Architect. Day-to-day supervision is therefore, apparently, left to the contractor and to the clerk of works representing the employer. On the other hand, the Architect has substantial powers under the main contract to enable him to enforce standards. He can call for verification that materials comply with specification. He can require that work be opened up for inspection. He can order the removal from site of defective materials. He can give instructions to the contractor on matters in respect of which he has contractual powers and, if these are ignored, the purchaser is entitled to employ someone else to give effect to the instruction and to deduct any costs incurred from moneys due to the contractor. These are not inconsiderable powers. In addition, the Architect's role in the certification of payments to the contractor gives him the power to refuse to certify work which fails to comply with specification.

All these powers would appear to imply a clear duty on the part of the Architect to make sure that the finished building complies in every respect with the specification. In practice, however, the position is ill-defined and it is not easy for a purchaser to obtain redress under contract from an architect in respect of a defective building. Given the nature of building work, this is not as unreasonable as it might at first sight appear. After all, a wise man does not accept responsibility for activities which are outwith his control. For an architect to accept full responsibility for quality it would be necessary for him or his representative to be present to observe, measure and test every single element of the building. This would clearly be impractical and indeed superfluous since the main contract obliges the contractor to comply with all the requirements of the specification and this obligation is not diminished by the issue of a certificate for payment.

Responsibilities of the Engineer

The sequence of activities undertaken by an Engineer appointed under the ACE Conditions of Engagement is not dissimilar to that of an Architect as defined in the RIBA form of agreement. On the other hand, the responsibilities he owes to his client (the purchaser) are more extensive and these are matched by greater powers to control work on the construction site.

The construction of a building is a reasonably predictable process. Only the part below ground, the foundation, is subject to unforeseeable events. When ground is excavated the unexpected can always happen, no matter how many exploratory investigations have been made. Likewise the behaviour of piles can never be predicted with certainty. For this reason, the engineer responsible for a building's foundation has to be prepared to respond to the unexpected and to adjust his plans in accordance with the ground conditions that are exposed. However, once out of the ground a builder is soon in a controlled environment entirely of his own making.

In civil engineering, the proportion of work subject to the whims of nature is generally higher than is the case with building works. Tunnels, for example, are never out of the ground. Earth dams and embankments are built from natural materials and methods of placing and compaction are subject to constant adjustment. Marine structures combine uncertain sea bed conditions with the unpredictable effects of winds, waves and tides. Collapse of such works, either during or after completion of construction, can have catastrophic effects on whole communities. There is thus a need for continuous and close involvement of the designers in the supervision of construction of civil engineering work so that informed and competent decisions can be made on the spot and without delay as the need arises.

Under the ACE Conditions of Engagement, the consulting engineer is expected to provide all the expert technical advice and skills which are normally required for the class of work being undertaken. The word 'normally' is important here, and there are various specialist services for which the Engineer will not be held responsible unless specially commissioned. These include architecture, although if the purchaser has also appointed an Architect, the Engineer will be expected to consult with him in connection with the architectural treatment of the work and to collaborate in the design of building works included in the main contract. Thus, whereas the Architect's role is to determine the concept and appearance of a building and thereafter to work with other consultants employed by the purchaser to produce the completed design, the Engineer is in a position both stronger and more onerous in that he alone is responsible to the purchaser for both concept and all other design work apart from any specialist or abnormal aspects which may have been excluded from his brief.

To ensure the necessary integration of the functions of design and construction, the Engineer has powers for controlling the site which exceed those of the Architect. For example, whereas the Architect's authority to give instructions to the contractor is limited to matters in respect of which he is expressly empowered under the contract to give instructions, the Engineer can give directions on any matter connected with the works, whether it is mentioned in the contract or not. The materials, plant and labour provided by the contractor, and the mode, manner and speed of construction have all to be to the Engineer's approval. He can call for details of the contractor's proposed construction methods and temporary works, and if he has good reason not to be satisfied with these he may reject them and require the contractor to submit new proposals.

The implementation of these and other powers requires personal representation on the site. The Engineer's contract with the purchaser stipulates that if, in his opinion, the nature of the works warrants engineering supervision of the site, the Client (or purchaser) 'shall not object to' the appointment of such suitably qualified persons as the Engineer may consider necessary. Persons so appointed are usually employees of the consulting engineer, although very occasionally and subject to the agreement and approval of the Engineer, they may be employed by the Purchaser. In either case they take instructions only from the Engineer. The person in charge of the Engineer's site team is known as the Engineer's Representative, and it is a requirement of the main contract that the contractor should be informed of his appointment and be given written details of the powers and responsibilities delegated to him.

It is of interest to compare the functions of the Engineer's Representative under the ICE Conditions of Contract with that of a clerk of works on a building project. Both occupy a key role in the management of quality. At one end of the spectrum, the role of the Engineer's Representative may be identical to that of a clerk of works as described on p. 21. On the other hand the Engineer may choose to delegate a significant proportion of his powers to his Representative who may, for example, be given authority to approve or reject work, to give instructions and directions to the Contractor, and to certify interim payments. He may not, however, relieve the contractor of any of his obligations under the contract, nor commit the purchaser to any extra payments or extensions of time. These are matters reserved for the personal attention of the Engineer.

Sub-contracts

Few main contractors these days undertake all the work involved in a contract on their own account. To do so would require that they maintain resources of men and equipment which would inevitably be under-utilized much of the time. Thus, subject to the provisions of the main contract, they prefer to award sub-contracts whereby particular elements of the work will be performed by others.

As mentioned on p. 19, most forms of contract prohibit sub-contracting without the permission of the Engineer or the Architect. There are good reasons why this should be so. For a start, if contractual obligations are carried out by persons who are not party to the main contract, it becomes more difficult for the Architect or Engineer to keep track of what is going on. Or it may be that the main contractor has only qualified to bid for the work because of his specialist capabilities which would be denied to the purchaser if some other less-specialized company were to do the work. On the other hand, there are times when the purchaser (acting perhaps on the advice of his Architect or Engineer) decides himself that parts of the work should be sub-contracted to persons or companies he has selected. This frequently applies in the case of specialists such as building services contractors. Sub-contracts of this kind are known as nominated sub-contracts, and the normal procedure is for the Architect or Engineer to obtain quotations for the work in question direct from prospective sub-contractors and to instruct the contractor to place an order with the chosen tenderer.

Sub-contracting work to others does not relieve the main contractor of any of his contractual obligations to the purchaser, and he normally remains fully responsible for the quality of the sub-contractor's work. Many contractors attempt to impose their own terms and conditions of contract on their sub-contractors and these sometimes contain one-sided provisions which place the sub-contractor at a disadvantage. On the other hand, some sub-contractors, particularly specialists, may refuse to accept work save on their own conditions. These problems may be avoided by using one of the standard forms of sub-contract designed for use with the standard main forms.

Since the purchaser has no contractual relationship with sub-contractors it is unwise for him, or for his Architect or Engineer, to communicate with them other than through, or in the presence of, the main contractor. On sites where there is a large number of sub-contractors performing a significant proportion of the total work effort, this can create major problems in the achievement of quality unless an adequate quality system is established and maintained by the main contractor. Particular problems may arise in the case of nominated sub-contractors where the main contractor is expected to accept responsibility for the quality of work of a sub-contractor in the selection of whom he took no part, and whose employment was imposed upon him.

Project management contracts

The traditional contractual arrangements outlined above have served their purpose admirably in the past, but there is a trend within the construction industry to seek other forms of contract more appropriate to today's conditions. The common feature of most alternative arrangements is the appointment by the purchaser of a management contractor to co-ordinate and manage both the design and construction phases of the project on his behalf. The management contractor is normally paid for his services by a fee which may be calculated in a number of different ways. He seldom undertakes construction work himself but may sometimes provide certain common services to the construction contractors such as canteens, site offices etc.

One of the causes of the trend towards alternative forms of contract is the growing proportion of work let to sub-contractors as the industry becomes more specialized. Another cause is the difficulty increasingly experienced by architects and consulting engineers in sustaining their often mutually conflicting roles as designers, supervisors and impartial adjudicators. Today's cut-throat competition leads contractors to rely increasingly for their profit margins on the successful prosecution of claims for disruption, unforeseen circumstances and variations. Many of these claims arise from the nature, adequacy or timing of design information. The Engineer or Architect responsible for design may then be charged with making impartial decisions on claims, some of which may arise because of his own errors or shortcomings. His position in such circumstances is barely tenable and as a consequence the role of the Engineer and Architect in the resolution of disputes is coming to be viewed with scepticism by a growing number of authorities on both sides of the industry. There are also moves to end the separation of responsibilities for design and construction.

Management contracts can offer partial solutions to many of these problems and are particularly effective for projects in which time is important and where there is a likelihood that designs will not be complete when a main contract would normally be let. They find especial favour in the process and offshore sectors of the industry where failure to complete on time can be very expensive, where the integration of design and construction is vital and where contractors have achieved a high degree of technical proficiency. They are also becoming increasingly popular for building projects, particularly those having a high building services content.

Figure 2.2 illustrates three alternative contractual and organizational arrangements for project management contracts. In all cases, the design organization retains full responsibility for design and for specifying the standards to be achieved. However the onus of achieving quality on site rests with the management contractor or construction manager who relies on his authority to approve payments for work to maintain control of the construction contractors. Although the designer's role in construction supervision is substantially reduced and he no longer holds the scales of

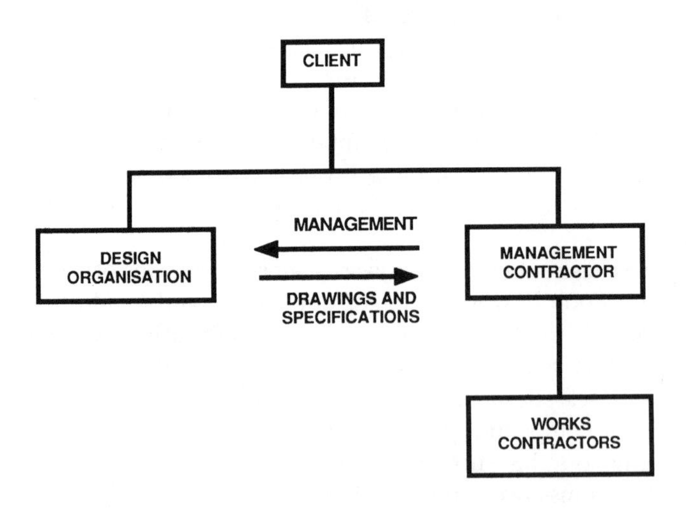

Figure 2.2 (a) The management contract.

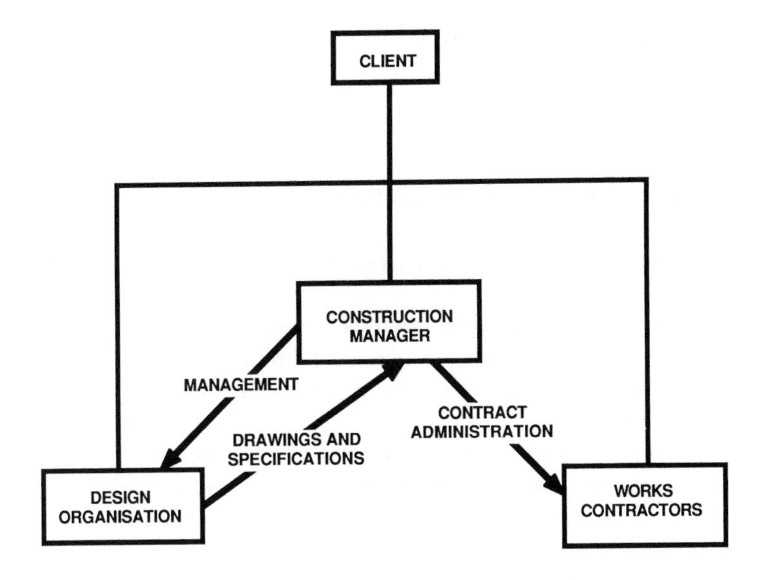

Figure 2.2 (b) The construction management contract.

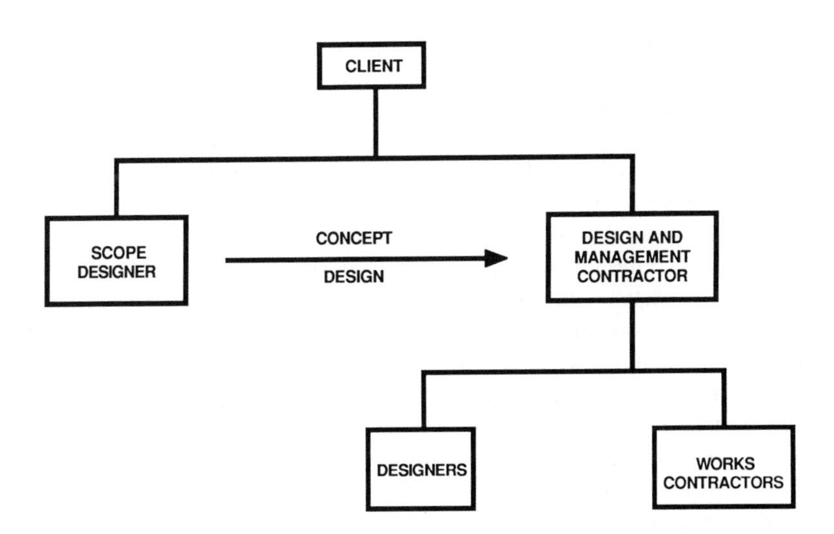

Figure 2.2 (c) The design and management contract.

justice, he retains the right of access to the site so that he can gain essential feedback of information and satisfy himself that his designs are being correctly interpreted.

House building for private sale

The provision of houses for private purchase represents a major sector of the construction market.

When projects are built under contract, construction does not commence until there is a firm and enforceable commitment between the purchaser and the supplier. This commitment requires that the purchaser should specify in precise detail exactly what he wants, that the supplier should name his price and that there should be a set of rules to ensure fair play between the parties. By contrast, most privately sold houses are built as speculations, that is to say the builder erects houses of his own design and at his own expense and then offers them for sale. If he has judged the market correctly he earns a profitable return on his outlay and his speculation has succeeded. If not, he loses money.

Speculative housing is therefore intrinsically different from construction under contract, indeed it has more in common with manufacturing and retailing. There is, however, a significant difference between the buying of a house and other private purchases in that it is usually the largest and most important commercial transaction which most people enter into during their lives. It introduces complex problems in respect of the legality and obligations of ownership and usually requires the raising of finance. It is also very difficult, if not impossible, for the average purchaser to satisfy himself that he is obtaining quality commensurate with the price he is paying. A good solicitor will resolve problems relating to title and ownership, and banks and building societies are available to provide finance, but what assurance can the purchaser have in respect of quality?

If the house is complete at the time of sale, it is open to the purchaser, or to a qualified building surveyor acting on his behalf, to examine the property and to satisfy himself as to its condition. This is a prudent action to take, but such examinations cannot be expected to be exhaustive. It is very difficult, for example, to check the condition of the foundations or the drains, or the thermal insulation. Nevertheless, if the opportunity is there it should be taken. A further source of assurance is provided by the fact that builders are obliged by law to comply with Building Regulations. The purpose of these is to protect the health and safety of the public and to prescribe minimum standards in respect of such matters as structural stability, fire protection, sanitation, internal heating and so on. All persons intending to erect buildings covered by the Regulations, and this includes most dwellings, are required to give notice of their intentions to the local authority and to deposit plans and specifications demonstrating compliance with the design requirements. After approval, further notices have to be given before and during construction so that local authority inspectors can visit the site and verify that buildings are constructed in accordance with approved drawings and specifications.

Additional protection is provided to the house purchaser by a body known as the National House Building Council (NHBC). This is a non-profit-making organization governed by nominees of various interested parties including building societies, consumer protection bodies, the professions and the building companies themselves. Its Chairman is appointed by the Secretary of State but it is otherwise independent of government and political parties. Most builders are registered with the Council and almost all new houses and apartments are covered by its arrangements. The NHBC publishes recommended specifications for design, workmanship and materials which amplify the more basic requirements of the Building Regulations. They employ field inspectors who make visits to sites both on a routine basis and also at random to ensure that specifications are being complied with. In so doing the Council acts as a third-party body undertaking audit and surveillance as described in Chapter 7 (p. 111), although it is not recognized by the National Accreditation Council for Certification Bodies.

The NHBC further protects the house purchaser by publishing a form of agreement and operating an insurance policy which registered builders are required to offer to prospective purchasers. These provide for the following:

1. Compensation to the purchaser in respect of losses incurred as a result of the builder's bankruptcy. This includes costs arising from failure to commence or complete construction to specification and reimbursement of any advance payments to the builder which cannot be recovered.
2. A two-year commitment by the builder to rectify, at his own expense and within a reasonable time, any defects and consequent damage to the property caused by a breach of the Council's requirements. In the event of disagreement between the purchaser and the builder, the NHBC will provide a conciliation service. If this fails, the Agreement provides for independent arbitration. If the builder then fails to carry out work found necessary by the arbitrator, the NHBC will meet the costs of having it done by someone else, subject to certain defined limits.
3. A further eight-year commitment (making ten years in all) during which the purchaser is protected against major damage caused by a defect in the structure or by subsidence, settlement and heave.

The NHBC agreement and insurance policies are lengthy legal documents, and the very brief summary of principles given above inevitably omits many important exclusions and conditions. They are undoubtedly an effective means of providing assurance to the house purchaser that he is receiving a building which is fit for its purpose and protecting him from financial disaster should things unfortunately go wrong. At the same time, they enable the NHBC to exert a powerful influence in the promotion of quality within the industry. Building societies are reluctant to grant mortgages on properties built by non-registered companies. Premiums paid by builders for registration and insurance are linked to their claims records. These factors, together with the NHBC's ultimate sanction of expulsion, provide a potent incentive to builders to meet NHBC standards and to accept their surveillance.

Contracts and the courts

There is a risk that the brevity of the foregoing summary of typical contractual arrangements used in the construction industry may have given a misleading impression of simplicity and clarity. Such unfortunately is seldom the case in practice. There are very many different forms of contract and these are frequently varied by purchasers to suit particular circumstances. Nevertheless, during the execution of the contract and through the ensuing maintenance period there is at least some common understanding of the responsibilities of the various parties involved and of the remedies available to each in the event of default by one or another. However, the products of construction are built for the long term and should outlast by many years the contractual obligations between the original purchasers, designers, material suppliers and contractors. When defects come to light after the expiry of the contractual maintenance period, who is liable, and to whom? Ask this question of a lawyer and he will smile with anticipation of a fortune in the offing. The English law on construction liability is confused and frequently incomprehensible, not just to laymen but also to the experts. Any attempt to unravel its many strands in a book such as this would be doomed to failure, but there are certain significant aspects which warrant attention.

The obligations of the parties to a contract cease when the contractor has completed all the work including maintenance, the requisite certificates have been issued and the purchaser has paid all amounts due. The contract is then said to have been performed and is 'discharged'. However, if latent defects are found before the expiry of a statutory period of limitation, it is held that the contract has not in fact been performed and is not discharged. The purchaser then regains his right to sue for breach of contract and to claim damages. This right applies not just to the contract between the purchaser and the contractor, but also to that between the purchaser and his professional advisers. The liabilities of the various parties should in theory be clear and unequivocal and capable of resolution according to the wording of the contract. But this is not always the case. The courts can, for example, impose 'implied terms' where contracts are not precise. They can also override clauses restricting liability if they judge them to be 'unreasonable'. Furthermore, it can be very difficult sometimes to decide whether a latent defect is the result of poor design and therefore the responsibility of the Architect or Engineer, or poor materials or workmanship resulting from shortcomings on the part of the contractor.

Under conventional arrangements, architects and consulting engineers are bound to exercise only reasonable care and skill in their work: they do not warrant that their designs will be fit for the purposes for which they may be intended. On the other hand, a contractor who accepts a commission to design and build a structure undertakes a greater obligation. In the absence of any express term to the contrary in the contract, the law will imply a term that the finished work will be reasonably fit for the purpose required. There are standard forms of contract for design and build work, and at least one of these expressly provides that the contractor should carry no greater a liability to the purchaser than would an architect or other professional designer acting independently under a separate contract with the purchaser. Under such arrangements, the dissatisfied purchaser can obtain redress only if he can prove negligence, and this may be difficult.

Whereas only those who are or were party to the contract can make contractual claims, any aggrieved person may exercise common law rights and sue under the law of tort. Tort may be defined as a civil wrong independent of contract, or as liability arising from a legal duty owed to persons generally. This widens the scope of possible claimants to include tenants of property, owners who were not original purchasers and, indeed, any other user who can show that he or she has suffered from negligence. It also exposes many more people and organizations to the risk of litigation: for example, local authorities have been held liable to purchasers of defective houses for negligence on the part of building inspectors.

Decisions in cases brought under tort are based on case law according to precedents set in previous similar cases. Thus the law develops from case to case and can adjust itself to correspond to currently accepted notions of what the ordinary man expects. While this provides a welcome degree of flexibility in legal decision making, it also makes it extremely difficult for anyone to predict the outcome of a particular case. To succeed in claims under tort, it is normally necessary to prove negligence. Negligence implies a breach of a duty of care. But just because things go wrong from time to time does not mean that somebody, somewhere, has been negligent. Errors of judgement are not necessarily proof of a lack of care. If they were, then a professional person would be liable whenever something went wrong and life for engineers and architects would become impossible. Lord Denning recognized this in one of his judgements: 'Whenever I give a judgement, and it is afterwards reversed in the House of Lords, is it to be said that I was negligent? Every one of us every day gives a judgement which is afterwards found to be wrong. It may be an error of judgement, but it is not negligent.'

Because of the uncertainties of proving negligence and the likelihood of each defending party blaming another, it is a common custom in cases of latent defects for plaintiffs to take action against all parties concerned in the hope that at least one will be found negligent. Since it is seldom that all the blame can be laid at one door, the courts often allocate liability on a proportional basis against two or more defendants according to the degree of responsibility which they are judged to carry. At first sight, this may appear a satisfactory form of rough justice. But what happens if, say, the builder has gone out of business, or the architect has died, or the structural engineer has no resources? The answer is that if a defendant is unable to pay his share the onus is redistributed to the others on a pro rata basis. In theory, therefore, if only one defendant has funds, he, and he alone, may have to bear the entire cost.

Claims for damages, whether in respect of tort or of breach of contract, are subject to statutory periods of limitation after which they become 'time-barred' and action cannot proceed. Since the life expectancy of the products of the construction industry is long, these periods of limitation can have considerable financial significance to purchasers, contractors and professionals and they have been the subject of much controversy and legal argument. In 1984, the Law Reform Committee made recommendations for the reform of statutory limitations in respect of latent damage in negligence cases. The standard period of limitation at that time was six years, but successful legal judgements on the time at which the period started to run had given rise to doubts as to how the law would be applied in particular cases. The aims of the Law Reform Committee's deliberations were:

1. To provide plaintiffs with a fair and sufficient opportunity to pursue their claims.
2. To protect potential defendants against stale claims.
3. To eliminate uncertainty in the law so far as may be possible.

Parliament considered the Law Reform Committee's recommendations and subsequently embodied the following in the Latent Damage Act 1986:

1. The period of limitation for actions for breach of contract remains at six years, or twelve years if under seal, from the date upon which the breach is held to have occurred. The latter date may be assumed to be the date of contract completion, or the date upon which the final certificate for payment was issued. However, if further work is subsequently carried

out, for example, if the architect gives advice on defects coming to light after completion, this may override the original date and the six-year period, as far as the architect is concerned, would then be held to run from the latest date on which professional advice was provided.

2. A similar six-year period applies to actions in tort. As in actions under contract, this period commences from the time that the damage occurs, which may be taken as the date of contract completion. Architects and consulting engineers may have a continuing liability similar to that described in 1 above.
3. Action in tort may also be commenced after the end of the six-year limitation period providing that it is brought within three years of the date on which a defect is discovered or becomes discoverable by the plaintiff. The three-year period may commence during the latter half of the six-year period, in which case it has the effect of extending the six-year period by an appropriate amount up to a maximum of nine years. Alternatively it may start after the expiry of the six-year period, subject only to the 'long stop' described in 4. below.
4. A 'long stop' period of fifteen years from the last date on which there 'occurred an act or omission alleged to constitute negligence.'

The provisions of the 1986 act now await interpretation in the courts. While some of the obscurities which existed before 1986 have been resolved, others have not and it remains to be seen whether the aim of the Law Reform Committee 'to avoid uncertainty in the law whenever possible' will be satisfied.

Contractors do not normally carry insurance against claims for latent defects, although policies providing such cover are available. The imposition of heavy damages may force a company into liquidation, but limited liability status will protect the assets of shareholders. Architects and engineers normally act as unlimited liability partnerships or as sole practitioners. Their exposure to risk is therefore severe since, if found negligent, they stand to lose all their assets and even after their death, recovery of claims can still be made against their estates. For this reason, most architects and engineers carry Professional Indemnity insurance covering them against claims by their clients. Premiums for Professional Indemnity insurance have risen steeply in recent years and can be as much as 5–10% of turnover.

So, what can purchasers, contractors, architects and engineers do to protect themselves from involvement in latent defect litigation?

1. Settle out of court. Latent defect cases are expensive and time consuming. All the parties concerned are likely to require legal representation and employ expert witnesses and cases last, on average, about three years. Fees are commonly quite disproportionate to the costs of the defects.

More than half of claimants lose their case and have to pay for legal costs as well as for rectifying the defects.
2. Make sure that the defects do not occur in the first place. This is by far the best remedy and one which is readily and freely available within the industry. The faults which give rise to latent defects litigation are usually very basic and arise because of ignorance, idleness or greed. They can best be prevented by technical competence and careful supervision backed up by a comprehensive and disciplined quality system.

Construction contracts as quality systems

At the start of this chapter, it was stated that traditional contractual arrangements provide the framework of a quality system comparable to those described in quality system standards. As will become apparent in the ensuing chapters, one of the main purposes of a quality system is to ensure that a purchaser's requirements are satisfied by his supplier. To achieve this a quality system for procurement will provide for:

1. Precise definition of the purchaser's requirements.
2. Selection of potential suppliers who can demonstrate both the means and the will to meet the requirements.
3. Surveillance of work in progress.
4. Verification, at source or after receipt, that the purchased products or services are in conformance with the specified requirements.

How do traditional arrangements for procuring construction work satisfy these criteria?

DEFINITION

The specifications and drawings incorporated in construction contracts are usually voluminous and comprehensive, but the amount of detail they give is subject to practical limitations. It is not possible, for example, for the drawings of a multi-storey building to specify the precise location of each brick or each nail. There comes a point at which the designer has to rely on the good sense of the builder to interpret what is needed and to follow customary good practice in carrying out the architect's instructions. Thus an architect might reasonably respond to a charge of defective design resulting in a damp or leaking building by saying that a competent builder should be aware of the need for proper damp-proof courses and other waterproofing measures and the absence of specific requirements for such measures on the drawings is no excuse for his failing to provide them. Such arguments are frequently upheld by the courts. However, the fact that litigation is being resorted to with increasing frequency is in itself evidence of an unsatisfactory system. The division of responsibilities for design and construction is a workable convention only if the boundaries are clearly defined and understood. This is frequently not the case, and the resulting confusion is the cause of many quality problems and consequent customer dissatisfaction.

From the purchaser's point of view, then, traditional arrangements do not always provide for precise definition of his requirements leading to his ultimate satisfaction. Fortunately, in most cases, the procedure of an architect or engineer producing concept drawings for formal approval by the purchaser and thereafter preparing detailed drawings and specifications provides a satisfactory outcome. But a purchaser who takes possession of a building or civil engineering structure and then finds that the designer has failed to understand and cater for his requirements has very few practical remedies, and this must be taken as indicative of a defective quality system.

SUPPLIER SELECTION

There are two principal 'suppliers' whose selection will influence the successful outcome of a construction project. The first of these is the architect or engineer who will design and supervise the works and the second is the contractor.

It has traditionally been considered unprofessional for architects and consulting engineers to advertise or otherwise tout for work and they do not normally enter into fee competition. A purchaser wishing to engage an architect or engineer will approach firms known to him or seek the advice of a professional organization such as the Association of Consulting Engineers or the Royal Institute of British Architects who will nominate members skilled and experienced in the particular field of work. The purchaser will interview potential contenders and make his selection according to his judgement of their professional capability.

The ACE and RIBA conditions of engagement stipulate that fees paid to architects and engineers will be in accordance with standard scales. These are of two kinds. Firstly there is a percentage of the total construction cost of the work which varies according to the size of the project (the higher the total cost, the lower the percentage) and the type of work (opera houses earn a bigger fee than farm buildings). Secondly there are fees charged on a time basis, normally for staff employed on site. It can thus be argued that the selection of architects and consulting engineers is made solely on their ability to satisfy the purchaser's requirements, unsullied by commercial competition. Given that the architect's or engineer's fee is generally a minor item in the total cost and that skills in design and supervision can bring benefits to the purchaser far in excess of their cost, there is a sound case in favour of such arrangements. On the other hand, the architect or engineer has little incentive to control cost inflation since his fees rise with the final bill. He is also subject to a conflict of interest when asked to adjudicate on a claim for extra payment by the contractor since a percentage of any sum agreed will accrue to himself.

Whereas architects and consulting engineers are usually chosen on the basis of their skill and experience, the conventional method of selecting contractors is by competitive tender. Award of work to the lowest bidder should in theory be the most effective way of minimizing the purchaser's expenditure. It is fair to contractors who expend effort and resources in tendering and who are entitled to know the basis upon which the successful tenderer will be chosen, and it provides a safeguard against corruption which is particularly important where public funds are involved. The danger in open competitive tendering is that the lowest bidder may be ignorant of the work for which he is tendering, and may not possess the skills and equipment needed to perform the contract to specification and programme. To avoid this, low bids deemed incompetent are sometimes disqualified and the work awarded to the next tenderer in line. This wastes effort and can create friction. A better procedure is to invite tenders only from a selected list of contractors drawn up on the basis of their previous records or some other process of pre-qualification.

Few individuals making purchases for their own use buy only the cheapest articles available. They prefer to weigh value against cost so as to optimize the benefits they receive. Suppliers of manufactured goods and those in service industries vary their specifications and prices to meet what they perceive to be the public demand and thus secure or improve their market share. Construction contractors tendering for work, however, do not have the option of varying specifications in order to offer purchasers better value for money. At best, they are permitted to submit alternative designs to achieve the same end-product as that determined by the purchaser's designers. The rationale for this arrangement is that the Architect or Engineer will already have optimized the value/cost ratio on the purchaser's behalf, therefore all that is needed from the tendering process is to establish who can meet the specification at lowest cost. The assumption is made that all bidders are equally certain to produce constructed works which do in fact conform with the drawings and specifications — after all, the Engineer's Representative or clerks of works are employed to ensure that they do. The weakness of this argument is that, however excellent the supervision, things do still sometimes go wrong. When they do, the purchaser may seek redress from the contractor. If he has been selected on the basis of lowest cost, not only may he be the least likely to perform, he is also the least likely to be able to provide adequate compensation.

The preceding paragraphs relate to traditional methods of supplier selection in the construction industry. The growth of management contracting and other arrangements which combine responsibilities for design and construction is bringing about a change of mood wherein many established customs are being questioned. Rules governing advertising by consultants have recently been relaxed, and fee competition for design work is now not uncommon. The trend is towards an extension of the principle of competition in the selection of all participants in the construction process, moderated by more intensive examination of the quality systems of potential contenders before commitment to a contract.

SURVEILLANCE AND VERIFICATION

Supplier surveillance and verification in construction are subject to many of the same conflicts and confusions as have been identified in the supplier selection process.

How does a purchaser, particularly one lacking technical expertise, monitor and check and verify the work of an architect or consulting engineer? In fact there is very little that can be done once work starts, and the purchaser is almost entirely dependent on the professionalism of his consultant and the incentive he is under to maintain a reputation in order to sustain a future flow of work. Likewise on site, the surveillance and verification carried out under the contract by the Engineer's Representative or by the Architect and clerks of works is far from failure-proof and suffers from a lack of clear definition of responsibilities. In practice neither professional designers nor contractors provide much documentary or other evidence of compliance with specification, it being accepted that the completion of the works is *ipso facto* proof that all checks have been made and satisfied. When documentation is supplied, it is often not the most relevant to the needs of the purchaser.

Conclusions

In conclusion, then, the conventional system for assuring customer satisfaction and compliance with specification can only partially satisfy our requirements for a quality system. The principal areas of concern are the lack of integration between the designers and builders and the diffusion of responsibility for the supervision of work on the construction site. It is a system founded upon traditions established at a time when the few skills needed could effectively be handed down from father to son or from journeyman to apprentice, when much of the work depended upon brawn and endurance rather than brainpower, and when the empirical design methods used allowed ample margins for error. To-day's circumstances are very different, and it is hardly surprising that the construction industry's reputation for quality should be low. Reform is overdue.

3

STANDARDS AND TERMINOLOGY

Brief references have already been made in this book to quality system standards, in particular DEF STAN 05-21: *Quality control system requirements for industry* and BS 5750:1987: *Quality systems*. It is now necessary to examine these and other standards in greater detail and to form a clear understanding of their subject matter and terminology. In the paragraphs which follow, certain passages from standards will be selected for analysis and discussion. Selection is not possible without omission, and for a full understanding of the standards there is no substitute for careful study of the original documents.

The standards most frequently encountered in construction work in the United Kingdom will now be described under two headings, 'general-purpose standards' and 'nuclear standards'. Both general-purpose and nuclear standards specify systems which will maintain and assure quality; the difference is that whereas the first category is oriented to the requirements of the market place, the second is aimed more at satisfying the statutory requirements imposed by regulatory authorities particularly in respect of safety. This results in a difference of emphasis, although the ground covered and the systems specified are not dissimilar.

General-purpose standards

Most standards in this category owe their origin to the AQAP series established for defence procurement purposes by the North Atlantic Treaty Organization in 1968. The AQAP series were adopted by the British Ministry of Defence in 1970 (DEF STAN 05–08) and, following the 1977 report by Sir Frederick Warner on 'Standards and specifications in the engineering industries', the first version of BS 5750 was published by the British Standards Institution in 1979 to rationalize what had become a proliferation of standards issued by various purchasing and third-party organizations. BS 5750:1979 formed the centre-piece of a series of standards issued by the British Standards Institution on a number of quality-related subjects including metrology, reliability, measurement and calibration. Its issue marked a watershed in the development of quality systems for general industrial and commercial use. Its provisions were followed closely by a number of equivalent European national standards and it provided a foundation for the international standards for quality systems issued in 1987 by the International Organization for Standardization (ISO).

The ISO standards for quality systems are known as the ISO 9000 series. This consists of three standards designed for contractual use, two guidance documents and a vocabulary of terms. The series was adopted by the European Committee for Standardization (CEN) and it now forms part of the national standards systems of the members of CEN, which includes the sixteen member states of the European Economic Community and the European Free Trade Area. It has also been incorporated into the American ANSI/ASQC series of standards. Of the major western industrial nations, only Canada has retained its own national quality standards. In the United Kingdom, the British Standards Institution reproduced the ISO 9000 series of standards as the 1987 version of BS 5750. The equivalent standards in other major national systems are listed in Table 3.1.

Thus BS 5750:1987 follows a philosophy generally similar to that of the 1979 issue, although there have been changes in detail and wording. Many of these arose from the elimination of military terminology which had made the standard difficult to interpret in a commercial environment. It is supported by BS 4778:1987 *Quality vocabulary*, Part 1 of which is identical with ISO 8402:1986 *Quality—Vocabulary* and provides definitions of 22 terms in common use in the quality context together with their equivalents in French and Russian.

As Table 3.1 shows, BS 5750 is issued in four parts:

Part 0 Principal concepts and applications.

Section 0.1 Guide to selection and use.

Section 0.2 Guide to quality management and quality system elements.

Part 1 Specification for design/development, production installation and servicing.

Part 2 Specification for production and installation.

Part 3 Specification for final inspection and test.

Table 3.1 Quality system standards

Internal Quality Management

Standards Body		Quality management and quality assurance standards – Guidelines for selection and use	Quality management and quality system elements – Guidelines
International	ISO	ISO 9000	ISO 9004
European	CEN	EN 29000	EN 29004
United Kingdom	BSI	BS 5750:Part 0 Section 0.1	BS 5750:Part 0 Section 0.2
USA	ANSI/ASQC	Q90	Q94
West Germany	DIN	ISO 9000	ISO 9004
France	NF	X 50–121	X 50–122

External Quality Assurance

Standards Body		Quality systems – Model for quality assurance in design/development, production, installation and servicing	Quality systems – Model for quality assurance in production and installation	Quality systems – Model for quality assurance in final inspection and test
International	ISO	ISO 9001	ISO 9002	ISO 9003
European	CEN	EN 29001	EN 29002	EN 29003
United Kingdom	BSI	BS 5750:Part 1	BS 5750:Part 2	BS 5750: Part 2
USA	ANSI/ASQC	Q91	Q92	Q93
West Germany	DIN	ISO 9001	ISO 9002	ISO 9003
France	NF	X 50–131	X50–132	X50–133

Part 0 is a guidance document. Section 0.1 outlines the principal concepts and offers guidelines for the selection and use of the appropriate part of the standard for different applications. Section 0.2 describes the basic elements of a quality system and advises on their development and implementation. Note that Section 0.1 gives guidance on obtaining 'external' assurance from suppliers or contractors, whereas Section 0.2 gives 'internal' advice on how to establish one's own system.

In contrast to Part 0, which can have no contractual status, Parts 1,2 and 3 are for use in contracts when a purchaser (or client) requires the supplier (or contractor) to operate a quality system which will demonstrate, or give assurance, that he is capable of controlling the work which is to be undertaken. The three parts each describe a model system and the intention is that the purchaser will be able to select a model which will be appropriate to the particular product or service he is buying. As may be judged from the titles of the three parts, the models are in decreasing order of complexity, and whereas Part 1 specifies 19 elements of a system, Part 3 has only 12. The following is offered as an interpretation of the three levels in the context of construction.

Part 1 is for use when detailed specifications are not available and the purchaser's requirements have yet to be established or can only be stated in terms of the performance to be achieved. The supplier is expected to develop the design and to control quality throughout all stages of the work. Part 1 would thus be appropriate to Project Management contracts in which one contractor undertakes total responsibility for a project including design, construction and commissioning. They would also be applicable to contracts for the shop-detailing, supply and erection of structural steel or for the design, fabrication and installation of a heating and ventilating system.

Part 2 applies when the requirements of the purchaser can be stated in terms of an established design and specification but where conformance to these requirements can be adequately established only by inspections and tests performed during manufacture or construction. Such specifications could thus be applied to typical construction contracts let against drawings and technical specifications supplied by or on behalf of the purchaser. Other applications could include the supply of reinforcing steel or pre-cast concrete units.

Part 3 is applicable to products of established design whose conformance with specification can be established by inspection or testing in their finished state. Examples of such products encountered in the construction industry include aggregates, window glass or sanitary ware.

These quality system standards are designed for a number of uses. An organization may, for example, decide to set up a quality system in accordance with a chosen standard in order to safeguard the quality of its work and to satisfy its own management's need for assurance that its customers are receiving the products or services which they have specified. Guidance on establishing such a system may be obtained from BS 5750 Part 0.2. However, the organization may well find that an important part of its system will be the procedures to be followed to gain assurance that its suppliers are not jeopardizing its efforts by delivering sub-standard materials or components. If so, Part 0.1 will advise on the selection of the appropriate system model (Part 1,2 or 3) against which potential suppliers' systems can be assessed or audited before a contract is entered into. The organization may then go further and decide to make compliance with the selected quality system standard by its suppliers a contractual requirement. Having implemented all these activities, the intention is that the organization will not only be more certain of its own and its suppliers' ability to perform, it will also be able in turn to provide similar assurance to its customers. It is well to be aware, however, that although they are designed for this purpose, the invocation of quality system standards in construction contracts is not without its pitfalls. These are discussed further in Chapter 10.

The principal system elements addressed in Parts 1,2 and 3 of BS 5750 are listed in Table 3.2. Each will be discussed later in this book under the appropriate chapter heading.

Nuclear standards

The system standard for nuclear construction in the United Kingdom is BS 5882 *Specification for a total quality assurance programme for nuclear installations*. This follows established international practice and is compatible with the International Atomic Energy Authority's Code of Practice 50-C-QA and the International Organization for Standardization's standard ISO 6215.

The contents of BS 5882 are scheduled in Table 3.2 alongside those of BS 5750. Reference to this table will show that the subject matter of BS 5882 differs only marginally from those of the level 1 or 2 general-purpose standards. The differences which do exist show a greater emphasis in the case of nuclear standards upon the verification aspects of quality management. The term 'nuclear installations' includes within its scope not just nuclear power generation, but other associated activities such as the handling and storage of spent fuel, fuel processing, waste processing and waste disposal. The standard covers the whole extent of these activities from concept design through to detailed design, procurement, manufacture, construction, commissioning, operation and eventual de-commissioning.

While the record of the nuclear industry is generally very good, the potential dangers it presents to the health and safety of the public are

Table 3.2 Subject matter of quality system standards

Subject	Clause number			
	BS 5750:Pt.1 (ISO 9001)	BS 5750:Pt.2 (ISO 9002)	BS 5750:Pt.3 (ISO 9003)	BS 5882 (ISO 6215)
Management responsibility and organization	4.1	4.1	4.1	2
Quality system/programme	4.2	4.2	4.2	1
Contract review	4.3	4.3	—	—
Design control	4.4	—	—	3
Document control	4.5	4.4	4.3	6
Procurement document control	—	—	—	4
Instructions, procedures and drawings	—	—	—	5
Purchasing	4.6	4.5	—	7
Purchaser supplied product	4.7	4.6	—	—
Product identification and traceability	4.8	4.7	4.4	8
Process control	4.9	4.8	—	9
Inspection, testing and surveillance	4.10	4.9	4.5	10,11
Inspection, measuring and test equipment	4.11	4.10	4.6	12
Inspection, test and operating status	4.12	4.11	4.7	14
Control of non-conforming products	4.13	4.12	4.8	15
Corrective action	4.14	4.13	—	16
Handling, storage, packaging and delivery	4.15	4.14	4.9	13
Quality records	4.16	4.15	4.10	17
Quality audits	4.17	4.16	—	18
Training	4.18	4.17	4.11	2
Servicing	4.19	—	—	—
Statistical techniques	4.20	4.18	4.12	—

frightening to many people. Although the risks may be small in statistical terms, such is the nature of the hazard that it raises fears in the public mind which, some might argue, are quite out of proportion when set against all the other dangers to which humanity is prone. Nevertheless, public disquiet is not to be denied and, for this reason, most countries exercise strict controls on the construction and operation of all nuclear facilities through the activities of regulatory authorities endowed with statutory powers. The relevant licensing authority in the United Kingdom is the Nuclear Installations Inspectorate (NII). Any prospective owner or operator of a nuclear facility must satisfy the NII on the arrangements for the assurance of quality, and as part of the licensing process must submit documentary evidence of such arrangements to NII for approval. In the United Kingdom, compliance with BS 5882 will normally be regarded as adequate for this purpose. It is the owner of a nuclear facility who carries the burden of satisfying the regulatory authorities, so it is to the owner that nuclear quality system standards are addressed. The owner is obliged to implement a quality system for the project as a whole and to ensure that all participants, including designers, contractors, sub-contractors and any other parties contributing to the work all abide by its requirements.

A nuclear facility is made up of a number of elements. Some of these, such as reactor vessels and containments, have a direct impact on plant safety. Other elements, the canteens or maintenance workshops for example, are no more hazardous than similar structures in non-nuclear plants. Most nuclear quality system standards recognize this diversity. BS 5882, for example, sets out principles which *must* be applied to all safety-related items and which *may* be applied to other activities which affect the successful operation of the installation. It acknowledges that 'items and services will differ in regard to relative safety, reliability and performance importance' and concludes 'it is possible to use various methods or levels of control and verification to assure adequate quality'. The onus is upon the owner to establish the relative importance of different parts of the work and, subject to the approval of the regulatory authority, to ensure that appropriate system standards (which may include the general-purpose standards referred to previously) are specified and adhered to.

Terminology

So far, this book has been couched in terms which, it is hoped, will be familiar to and understood by most English-speaking persons. In writing on the subject of quality management, this is no mean feat, since it is its misfortune to have become so enmeshed in jargon that its essentially simple principles are at risk of becoming lost in a fog of obscurity. This may be due to the transatlantic origin of much of the early work on the subject. It may also be because of the need to comply with national and international standards which have perforce to be drafted and agreed to by committees.

Unlike most subjects of specifications and standards, quality management is an abstract concept. Whereas it is fairly simple to write precisely about tangible things, it is much more difficult to define abstractions with any degree of precision. Reference has has already been made to BS 4778, which provides definitions of terms used in related standards. BS 5882 also includes a section which defines the terms used therein. Not only is the jargon of quality management extensive, to add to the confusion different standards have in the past interpreted words in different ways. With the publication and adoption of the ISO 9000 series of standards in 1987, many of the differences in interpretation were resolved, but a few still remain. It is thus necessary to be aware not only that 'Many of the terms used ... have specific meanings rather than the generic definitions used in dictionaries' (BS 4778), but that a definition used in one standard may not necessarily apply in others.

The purpose of labouring this point is to warn the reader of the existence of the code of special phrases and meanings which comprise so much of the written matter associated with quality management. When used between experts, such a code can be useful if it enables ideas to be exchanged more rapidly and with greater precision than can be achieved with ordinary language. Many groups of experts use jargon ostensibly for this purpose — the legal and medical professions are two examples. The practice becomes reprehensible when it is used to confuse the layman or to confer a false mantle of superiority upon the self-styled expert. The use of jargon in this book will be kept to a minimum. When used, its meaning will be explained. Appendix A gives the principal definitions of the more important terms, but most readers will find that this adds to, rather than subtracts from, the confusion. The following paragraphs will attempt to present some of the terms in a more 'user-friendly' fashion.

QUALITY, PRODUCTS AND SERVICES

The following is a dictionary definition of the word 'quality'.

Quality
1. Degree of excellence, relative nature or kind or character.
2. Faculty, skill, accomplishment, characteristic trait, mental or moral attribute.
3. High-rank or social standing.
4. (Of proposition) being affirmative or negative.
5. (Of sound, voice, etc.) distinctive character from pitch and loudness, timbre.

Compare this with the British Standard definition:

Quality The totality of features and characteristics of a product or service that bear upon its ability to satisfy stated or implied needs (BS 4778 *Quality Vocabulary* Part 1).

The latter definition may be pedantic, but with careful reading its meaning becomes clear and is, of course, that ascribed to the word at the beginning of Chapter 1. But what is meant by the word 'needs'? BS 4778 offers the following notes in clarification:

'1. In a contractual environment, needs are specified, whereas in other environments, implied needs should be identified and defined.
2. In many instances, needs can change with time; this implies periodic revision of specifications.
3. Needs are usually translated into features and characteristics with specified criteria. Needs may include aspects of usability, safety, availability, reliability, maintainability, economics and environment.'

Under conventional contractual arrangements in the construction industry, it is the role of the design professionals to identify the purchaser's stated or implied needs and to translate these into drawings and specifications suitable for incorporation into a contract. It could thus be argued that for a contractor, 'quality' need mean no more than 'compliance with the drawings and specifications'. The contractor builds what he is contracted to build, and he cannot be expected to be aware of all the stated and implied needs of his clients, let alone those of the client's tenants or of subsequent owners. Unfortunately, however, as was briefly discussed in Chapter 2, life is seldom this simple and the precise legal liabilities of contractors, engineers and architects in respect of satisfying clients' needs are likely to remain obscure. These are muddy waters, and while 'compliance with drawings and specifications' may in some circumstances be a satisfactory interpretation of the word 'quality', its validity is only partial. Both the definition and the law recognize that implied needs have also to be taken into account.

The BS 4778 definition of quality relates to 'products and services'. The standard explains that these may be:

'— the result of activities or processes (tangible product; intangible product, such as a service, a computer program, a design, directions for use), or
— an activity or process (such as the provision of a service or the execution of a production process).'

Quality systems therefore are not confined to processes where there is a tangible end-product. The product of a garbage disposal organization is well-swept streets and empty bins. The product of a management consultant is sound advice. The product of an air-stewardess is service with a smile. All of these activities can be done well or done badly. The measure of their quality lies in the perceptions of those for whose benefit they are carried out.

MANAGEMENT, SYSTEMS AND ASSURANCE

The concepts of quality management and the application of quality systems have already been discussed at some length in this book. The term 'quality assurance', however, while it has been quoted in the titles of standards, has not been referred to in the text. Its omission has been deliberate as it is an expression which is frequently misunderstood and often misused in writings and discussions on quality. This nettle must now be grasped.

The definitions given below are from BS 4778 *Quality vocabulary*. They are chosen because they are reasonably easy to understand and they relate satisfactorily one with another.

Quality policy The overall quality intentions and directions of an organization as regards quality, as formally expressed by top management.

Quality management That aspect of the overall management function that determines and implements the quality policy.

Quality system The organizational structure, responsibilities, procedures, processes and resources for implementing quality management.

Quality assurance All those planned and systematic actions necessary to provide adequate confidence that a product or service will satisfy given requirements for quality.

Quality control The operational techniques and activities that are used to fulfil requirements for quality.

Thus quality ***management*** embraces all the actions an organization takes to achieve its quality ***policy***. Some of these actions may be unpremeditated and unsystematic, perhaps in reacting to events as they unfold, but most will follow organized routines and procedures established in advance. The latter form the quality ***system***. Such a system must of necessity be made up of a number of elements and these elements are identified and described in quality system standards. Some of these elements will provide ***quality control*** by eliminating non-conformance. Others will supply verification, or ***assurance***, that standards have been met — an assurance which may be made available to management, to the customer, to regulatory authorities, or to all three.

Clarification of these concepts is provided in BS 5750 Part 0.1. Clause 4 identifies three policy objectives which an organization should seek to accomplish with regard to quality:

'(a) The organization should achieve and sustain the quality of the product or service produced so as to meet continually the purchaser's stated or implied needs.
(b) The organization should provide confidence to its own management that the intended quality is being achieved and sustained.
(c) The organization should provide confidence to the purchaser that the intended quality is being, or will be, achieved in the delivered product or service provided. When contractually required, this provision of confidence may involve agreed demonstration requirements.'

So, quality systems have both to control quality and to assure it. They incorporate activities which provide operational controls, such as tests and inspections, and those which provide assurance, such as documentation and audits. Some activities provide both — tests and inspections carried out for operational purposes can also supply verification, and vice versa. The boundaries between control activities and assurance activities within a system are often shifting and difficult to identify and define; indeed, at any one time individuals may be performing both control activities and assurance activities. The existence of an effective system of quality control can, by itself, provide a potent form of quality assurance. Conversely, in the absence of an effective control system, no-one can have confidence in the product. The different ways in which systems may be organized to accommodate the needs for control and assurance are discussed in Chapter 4.

Manuals, plans and programmes

These words are used in quality system standards to identify the principal documents developed by management to describe and implement their quality systems. Some of these documents will be for the benefit of staff who have to put the systems into effect. Others will be prepared for issue to clients for assurance purposes, either before or after the signing of a contract. Some documents may serve both purposes. The standards listed in Table 3.1 and their accompanying vocabulary and guidance documents unfortunately are not consistent in their nomenclature, but careful study reveals that they do agree on the types of document needed. These may be summarized as follows:

Category 1 A document which states the quality policy of an organization as a whole and describes the system established for its implementation.

Category 2 A document which gives corporate instructions on the operational procedures to be followed to ensure product quality.

4

THE QUALITY SYSTEM

Management responsibility

Very few organizations are without systems of any kind. In small businesses where the person in charge can observe the activities of all employees and can personally check that no defective goods leave the premises, the system can be very simple indeed. However, as organizations become larger and more complex, there comes a point at which it is impossible for one person to undertake personal supervision of all that goes on and authority has to be delegated. At this point, the sensible manager will publish the rules which have to be observed and the procedures which are to be followed by members of staff exercising authority on his behalf. Some of these rules and procedures will relate to administrative and commercial matters, others will define how customer satisfaction with the products or services of the organization is to be assured. The latter category will form the kernel of the quality system.

But first it is necessary that management should define its policy and establish its objectives. Policy is the direction in which it wishes the organization to move. Objectives are the actions to be taken in order that it will do so. A quality system can function effectively only if it is part of an overall management system established to achieve stated objectives in accordance with a defined policy. Quality policy must rank alongside marketing policy, commercial policy, employee relations policy, and so on. Each is dependent upon the other. The topic of management objectives is further discussed in Chapter 11. At this point it is sufficient to note that a policy confined to the maximization of profit cannot be a satisfactory basis for a quality system. While profit is an important component in any organization's set of objectives, management must make it clear that profit is to be achieved by satisfying the customer, not by deceiving him.

Quality system standards require that responsibilities, authorities and inter relationships should be clearly defined by management. People with delegated responsibility for quality must have the freedom and authority to stop and reject work which is sub-standard, and to take action to prevent repetition. Staff engaged on inspection, testing and other verification are required to be properly trained.

Having established the system, management has a duty to make sure that everyone in the organization knows how it works and is aware of his or her personal role. To this end, systems should be documented and this is dealt with in Chapter 5.

Organization

A wide diversity of opinion prevails on the subject of organizational structures for the effective management of quality. This diversity reflects the differences in the processes, methods and cultural backgrounds of the various industries and companies concerned.

At one end of the spectrum, there are the comprehensive systems operated by multi-disciplinary companies which design their own products, purchase raw materials or components from others, carry out manufacture or assembly and then deliver the final product to the customer. By contrast, companies which supply standard products which can be adequately tested in their finished state will operate much simpler systems. As discussed in Chapter 3, these variations are reflected in the three levels of detail contained within the general purpose quality system standards. The construction industry contains organizations in all three of the categories for which these different levels are designed. The most comprehensive systems will be those developed by the larger multi-disciplinary contractors and similar consultancy practices, and it is with these organizations in mind that this chapter has been written. It is hoped that this will not deter the reader whose association is with smaller and less complex concerns. The principles of quality management are not dependent on size or complexity and the requirements of a small contractor or consultancy will differ little from those of sub-divisions or subsidiaries of larger companies or firms.

There are few formally established quality systems operating within construction contractors and consultancies, although most have written or unwritten procedures intended to serve the same objectives. Because of this lack of proven experience it is unrealistic to attempt to lay down precise guidelines. Nevertheless, it is both possible and reasonable to examine practice within industries where the principles of quality management have taken root, and extrapolate from these observations to identify principles which are likely to succeed in the construction industry.

In Chapter 3, it was established that a quality system comprises the management processes and resources assembled and implemented to achieve the organization's quality policy. BS 5750: Part 0 emphasizes the link between system and objectives:

> 'The quality system should only be as comprehensive as needed to meet the quality objectives.'
> (Note to paragraph 3.3, Section 0.1)
> 'A quality management system should be developed and implemented for the purpose of accomplishing the objectives set out in a company's quality policies.'
> (Paragraph 0.2, Section 0.2)

These quotations establish that the first objective in establishing a quality system should be to satisfy the internal needs of the organization. It follows therefore that it should be cost-effective, compatible with accepted good practice within the particular industry concerned, and beneficial to the organization. Part 0 of BS 5750, however, is only advisory. In contrast, Parts 1,2 and 3 of the standard are contractual, and they lay down requirements to be imposed by the purchaser on the supplier. Conflict may arise between what the purchaser wants in order to satisfy his quality policy and what the supplier finds appropriate for his own objectives. These conflicts are discussed further in Chapter 10. For the time being, let us concentrate on systems established for internal objectives. Most of these tend to fall into one of two general categories. These may be identified as the 'centralized' or 'de-centralized' concepts. Let us look at these in turn.

CENTRALIZED SYSTEMS

Centralized systems lay stress on the practice of quality control or 'the operational techniques and activities that are used to fulfil requirements for quality'. These obviously vary from one application to another. However, they will almost certainly include taking samples at various stages, comparing them with specified requirements and rejecting items which do not comply. Under a centralized system, these operations will be the responsibility of a quality control department with its own management hierarchy independent of production departments. In a large organization the quality control department may include technical experts such as materials engineers, metallurgists, physicists and computer programmers, as well as inspectors and testing operatives. They will have access to the works to take samples as required and will operate the laboratories in which testing is done.

Such centralized organizations can be very powerful. They provide both control and assurance of quality (QA/QC). They have the advantage that the authority and independence of the quality manager and his staff are clearly established and they can be effectively insulated from commercial pressures which might compromise their judgement. The disadvantage is that quality control departments tend to grow into separate empires working in parallel with but in isolation from those responsible for production. While they can be very effective in rejecting defective work, their isolation precludes them from participating in the planning and organization of procedures for the prevention of defects. Furthermore, in attempting to achieve and maintain the technical initiative they are prone to duplicate, if not outdo, the skills and qualifications of the production teams. Centralized systems therefore tend to be both expensive and productive of friction.

DE-CENTRALIZED SYSTEMS

The difference between centralized and **de-centralized** systems is that in the latter the responsibility for controlling quality is placed firmly on the shoulders of those actually doing the work. This follows the principle that production management has a duty to make things which comply with specification, a duty which it should not be permitted to relinquish or to share with others. It will be noted that this principle is in line with the concepts of quality management described on p. 5.

In a de-centralized system, production managers are required to develop plans, procedures and routines for inspection and testing which will ensure that work is done properly. Inspection and testing is then carried out mainly by staff within the production hierarchy in accordance with strictly defined and documented quality control programmes. With such arrangements, it requires little imagination to visualize that occasions will arise when an inspector's integrity will become stretched to the limit by the knowledge that his superiors will be displeased by a rejection which will disrupt production schedules. Protection against such pressure is provided by an independent quality assurance manager who has powers to approve or reject quality plans and to supervise their implementation by surveillance, random checks, examination of documentation and formal audit.

It may well be argued at this stage that in the construction industry we already have the bones of a de-centralized quality system. Reference to p. 19 will show that the role on site of the Engineer's Representative under the ICE Conditions of Contract or of the Architect under the JCT Standard Form of Building Contract is very similar to that of a quality assurance manager as defined above. They carry out surveillance, they do random checks, they examine documentation and, although the term may be unfamiliar to most of them, they audit the quality of work. In so doing, they supply verification to the purchaser that the work done on site complies with the specification. But is this enough? One has to conclude that it is not. The quality assurance activities of a client's representative on the site relate only to one project. They cannot provide the long-term continuity of assurance required from a contractor's quality system. Neither do they provide assurance in respect of design, either to the client or to the designer.

BS 5750 requires the appointment of a 'management representative' who, regardless of any other duties should be responsible for maintaining the system. In a large organization, the 'management representative' is likely to be a full-time post and the person concerned may carry the title 'quality director' or 'quality manager'. However, the standard does not make this mandatory, and part-time appointments are acceptable, although a part-time quality manager would not be able to carry out independent system checks in respect of work for which he is responsible. The role of the management representative will depend on the type of quality system adopted. The manager of a de-centralized system is likely to have a smaller permanent staff at his disposal and his operations will therefore be less expensive than those of a manager of a centralized system. Also his activities reinforce the responsibilities of the production team rather than detract from them. They thus serve to encourage a more serious and constructive attitude towards quality.

Opponents of de-centralized systems would respond that such benefits are illusory and that harsh experience teaches that to expect production teams to control quality is tantamount to commissioning Satan to prevent sin. Nevertheless, de-centralized systems do function effectively, and are compatible with quality system standards.

Group management structures

Most major contracting companies and the larger consultancies operate as groupings of subsidiary companies or divisions, each of which concentrates on a particular market sector. In the case of contractors, each of the subsidiary companies will have its own directors who will have autonomy within their company, subject to general policies and financial targets set by the board of directors and chief executive of the group. Typically, the chief executives of the subsidiary companies will also be directors of the main board of the group so that they can answer for their companies' performance and have a voice in the overall determination of policy. Likewise, in consultancies, each division will normally be under the control of a partner who will be responsible for the work of his division and who will participate with the senior partner and other partners in policy making for the firm as a whole. The paragraphs which follow are written in the context of the contracting side of the industry, but it is likely that the principles described will be equally applicable to the larger consultancies.

Suppose the directors or partners of a group decide to establish a group-wide policy in respect of quality management. There is a number of good reasons why such a decision should be made. One reason is that many large multi-disciplinary projects will require input from more than one subsidiary company or division and in these circumstances a common approach by each is desirable. Another reason would be to provide assurance to the group's management that individual companies or divisions are not jeopardising the group's long-term future by exposing it to loss of reputation or possible claims for compensation. A final reason, probably the best, is that the group wishes to reduce the costs of doing things badly so that they have to be done over again. Let us presume that a policy decision is made to establish a quality system designed to satisfy these objectives.

Should the group's quality system be 'centralized' or 'de-centralized'? A centralized system would require the establishment of a quality control team to be responsible for the inspection, testing and release of all the group's products and services. Few would disagree that such a system, while possibly suitable for a factory, would have little prospect of success in a construction organization carrying out a wide diversity of work in a multiplicity of locations and operating under a variety of contractual arrangements. Clearly a de-centralized system, emphasizing assurance rather than control, would be more likely to succeed. Such an arrangement would require each subsidiary company or division to develop its own quality system suitable for the market in which it operates, subject to approval and audit at group level. In this context, 'audit' has a special meaning which will become apparent later.

To give effect to this decision, the chief executive of the group, or the senior partner, would take the following actions:

1. Inform all sectors of the organization that a decision has been made to establish a group-wide quality system, explain its general nature and require that all staff co-operate in its implementation.
2. Appoint a director or manager as 'management representative' to initiate and monitor the operation of the system.

The initial announcement is an important document and requires careful thought. It should be confined to general principles and not go into detail. Most important, it should establish the fact that the quality system and those charged with its implementation have the whole-hearted support of top management. Many people will see the system as a threat to their own interests since it will cast light on matters normally kept hidden and may bring about change. It will therefore meet resistance, and unless those in charge have a clear mandate from the top, their efforts will be frustrated and the system will fail.

The success of the quality system will depend heavily on the personal qualities of the quality manager. It is a task which requires tact, integrity, persuasiveness and patience as well as a broad technical understanding. His or her location within an organization needs to be established with care. Impartiality and credibility demand a position which is free from immediate commercial pressures, and the need to command respect requires a status which will indicate that the holder has the confidence and support of the chief executive. A quality manager cannot function effectively without the understanding and support of all senior management, including production management. To this end, it is essential that he should have access to the decision makers of the organization he serves and be in a position which affords equality of status with those upon whose co-operation he depends.

In manufacturing industry it is common for the quality manager to be appointed a director of the company. This may also be the case with a major contracting group, although it is more likely that the role would be combined with other technical functions and represented at board level by a group technical director. In the latter case it is probable that the technical director would delegate the day-to-day implementation of the quality system to a full time quality assurance manager, whose functions might be listed as follows:

1. Preparation of a group quality manual (see Chapter 5).
2. Preparation of model quality system documentation.
3. Advice to operating companies or divisions on systems and documentation.
4. Periodic auditing of functioning of quality systems in operating companies or divisions.
5. Provision of an independent functional reporting line for quality management staff in operating companies or divisions.
6. Presentations to clients of group companies.
7. Co-ordination of recruitment and training of quality assurance staff throughout the group.
8. Preparation of periodic reports to the Group Technical Director for board presentation.

To fulfil these functions the group quality assurance manager would require a small staff, comprising perhaps a secretary, a deputy to assist in auditing and to stand in for him in his absence, and possibly one other to be available to assist operating companies with particular projects. Small is beautiful in this context, not only to minimize expenditure but also to refute those who might claim that formalized quality systems inevitably create large and unnecessary bureaucracies. Likewise, the group quality manual can be a slim document. It will state the group's quality policy, describe the organization at group level for its implementation and set out guidelines for the establishment of quality systems in operating companies. It is not likely to include or refer to standing instructions, since these are best promulgated at operating company level.

Company management structures

Let us now consider the quality systems of operating companies. In major groupings these are likely to be widely diversified. Their activities may include, for example, housebuilding, property development, civil engineering, commercial and industrial building, petro-chemicals, laboratories, management contracting, international contracting, and so on. To maintain a common thread in the midst of this variety, the first step is to issue an instruction requiring each company to appoint a director or senior manager to take responsibility for the establishment of a company quality system in accordance with group policy. The person appointed should be directly responsible to the chief executive of the operating company and have the necessary delegated authority and freedom from production pressures to enable him to discharge his duties effectively. This person would be the principal point of contact with the group quality assurance manager and his duties would typically include:

1. Plan, implement and maintain a company quality system.
2. Compile, update and issue a company quality manual.
3. Evaluate and approve quality systems and manuals of divisions and projects.
4. Provide an independent functional reporting line for quality staff in divisions or on projects.
5. Plan and direct internal quality audits.
6. Represent the company when it is audited by purchasers or third parties.
7. Liaise with the group quality assurance manager and respond to group quality audits.
8. Report to the company chief executive on quality matters.

Typical reporting lines and responsibilities of group and company quality directors and managers are illustrated in Figure 4.1. This indicates that the quality director is of equal status with the construction directors who control the company's projects and to whom all site staff are ultimately responsible. He is also of equal status to the directors responsible for the technical, commercial and administration functions. Although the quality director has overall custody of the quality system, the standing instructions (or procedures) which provide for its implementation have to be prepared and enforced by the appropriate line management and functional directors.

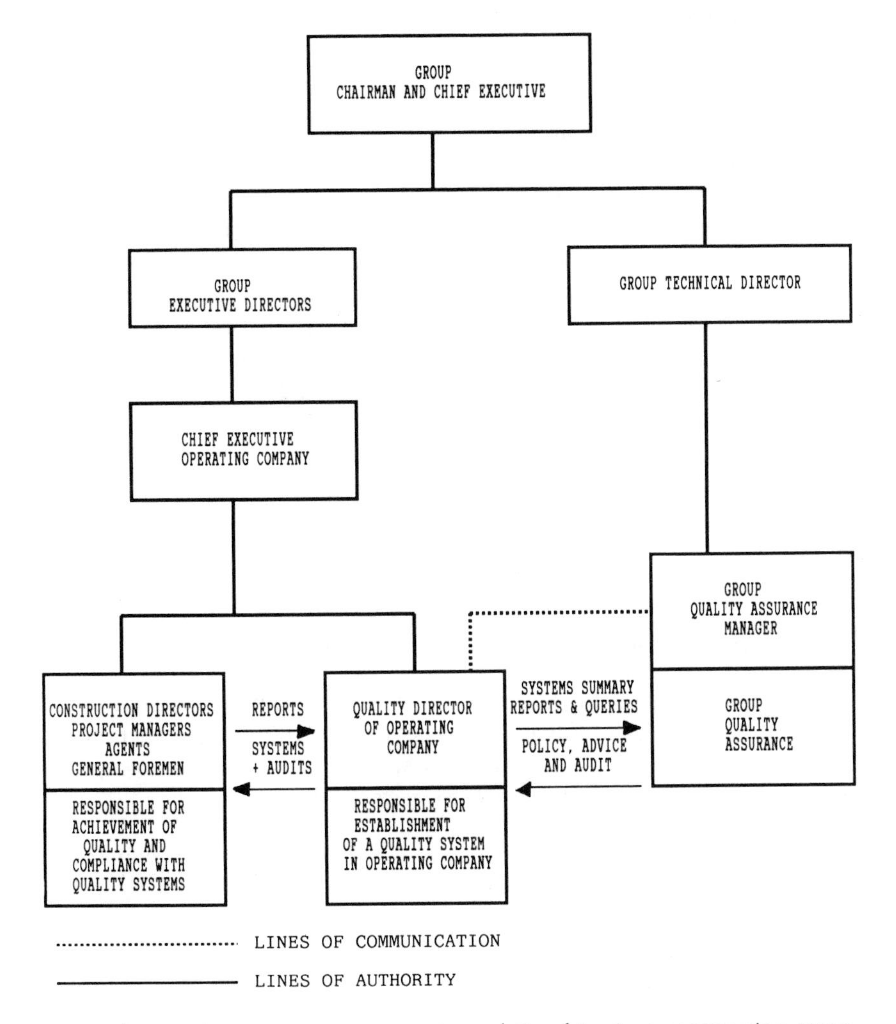

Figure 4.1 Quality management reporting relationships in a construction group.

They know best what is to be done and they, and only they, can exert the discipline necessary to ensure that instructions are adhered to.

Note that the director responsible for quality assurance has a direct line of communication to the group quality assurance manager, and through him to the group technical director and the group main board of directors. This line establishes the group quality assurance manager's right to request information and to carry out audits. Its presence also serves to discourage attempts to compromise the integrity of the company quality director.

Project management structures

The final tier in our management structure is the individual contract or site. These may vary in size from those supervised by just one or two full-time staff, to multi-million pound projects requiring a staff numbered in hundreds. The larger projects are likely to warrant a resident quality assurance management team implementing a project quality system, but this would not be economic for smaller projects if they can be adequately serviced on a visiting basis either by the company quality assurance manager himself or perhaps by a regional or divisional representative.

A typical organization chart showing the various management responsibilities on a major project is given in Figure 4.2. Let us consider the management posts identified on this chart.

The **Project Manager** would be appointed by, and would report to, the construction director responsible for the relevant market sector. His task would be to execute the project to specifications, budget and timescale. To achieve these objectives, he would have to delegate, and the chart indicates four departmental heads answerable to the project manager for the respective functions under their control. In making these appointments, the project manager should observe the two golden rules of delegation:

1. Delegation of duties from a delegator to a subordinate does not detract from the delegator's ultimate responsibility for the duties concerned.
2. Duties should only be delegated to subordinates who are competent, trained and experienced enough to perform them.

Each of the four departmental heads on our chart would be of equal status and all would be expected to communicate and co-operate with each other to bring the project to a successful conclusion. The project quality plan would contain a brief description of their respective duties.

The **Construction Manager** would be accountable to the project manager for the performance of work on site. The work would be as defined in the drawings and specifications and it would be the construction manager's duty to ensure that the finished product is performed within the permitted tolerances. The tests and checks which are necessary to achieve and verify compliance with specification are an inseparable part of the construction process and it therefore falls to the construction manager to organize the work in such a way that the requisite resources and time are made available for the tests and inspections which have to be carried out.

The **Commercial Manager**'s participation in the project quality system is likely to include the selection and control of sub-contractors and the purchasing of bought-in materials. He would also be responsible for stores managers and through them for the checking of materials on arrival at the site, for storage and security, and for final issue for use.

The **Chief Engineer** would be accountable to the project manager for the specific technical matters listed and he would also act as functional head of all engineers engaged on the project. He would be responsible for the overall planning of the work and for temporary works design. He would

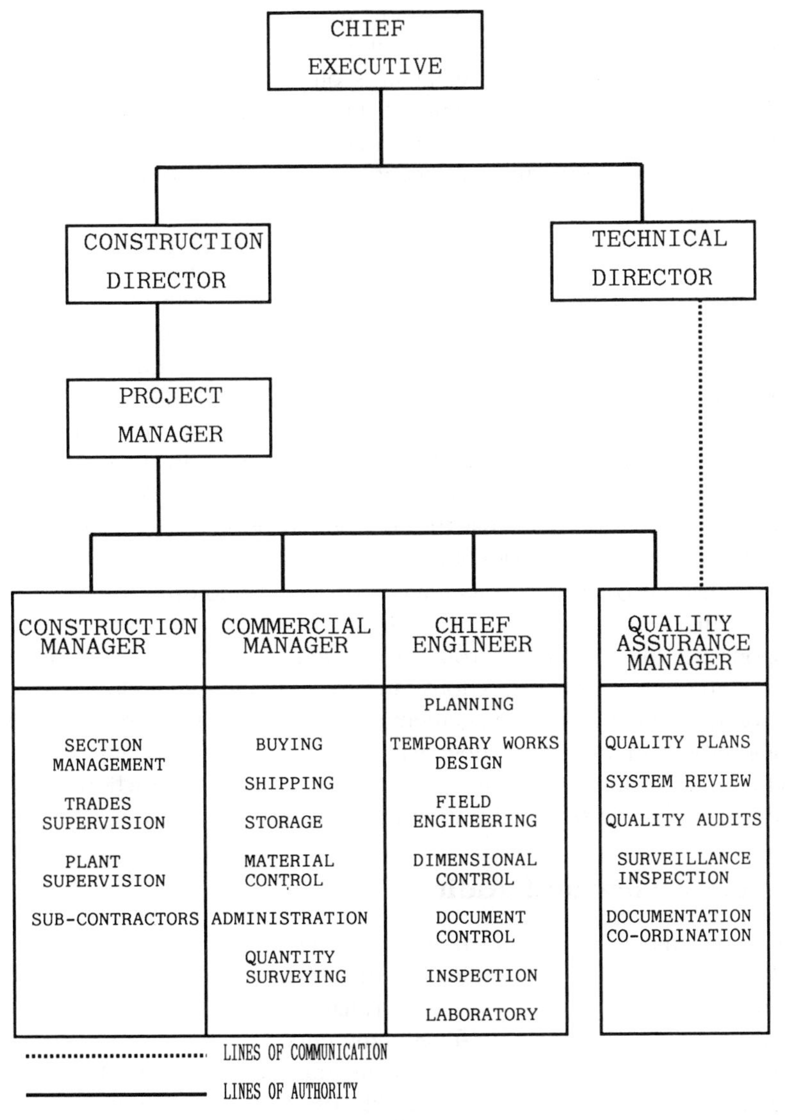

Figure 4.2 Typical project organization chart.

institute systems for the receipt and issue to site of drawings and other technical documents. Inspection staff engaged in the surveillance of work carried out in suppliers' works would answer to him, as would specialist inspectors for site operations such as the non-destructive testing of metalwork. Site laboratories would also be under his control.

The **Quality Assurance Manager** would occupy a position on the site analogous to that of a company quality assurance manager within the corporate structure. He would report on a day-to-day basis to the project manager who would delegate to him the task of preparing a project quality plan in accordance with the company's quality policy and in compliance with any contractual requirements of the purchaser. The project manager would approve the project quality plan and thereafter ensure that its requirements are correctly implemented by all personnel on the project.

Although reporting on the site to the project manager, the project quality assurance manager would also have access to the company quality assurance manager and through him to the board of directors. He would be responsible for managing the site quality assurance team which would be independent of other functional departments. The role of the quality assurance team would be to provide assurance that the requirements of the purchaser's contract specification have been correctly identified and complied with. It would verify that the project quality plan is effective and being complied with, that work is being done as instructed, that the required tests and inspections are being carried out, that documentation is complete and factual and that the appropriate levels of management are being informed of any shortcomings.

Note that a project quality assurance manager would discharge the role of 'management representative' as defined on p. 57. He would not be responsible for quality control. This, together with the preparation of procedures and work instructions, and the training and indoctrination of personnel would be line management responsibilities. These responsibilities would also embrace the supervision of sub-contractors and suppliers, including the approval of their procedures and work instructions, and the execution of inspection and testing programmes.

System review and audit

All management systems need to be regularly examined if they are to continue to be effective. Quality systems are no different, indeed they are particularly prone to deteriorate into ritual unless systematic measures are taken to ensure that they are functioning effectively and are responsive to current needs. In the jargon of quality assurance these systematic measures are known as 'system review' and 'audit'. The following definitions of these terms are from BS 4778:

Quality system review A formal evaluation by top management of the status and adequacy of the quality system in relation to quality policy and new objectives resulting from changing circumstances.

Quality audit A systematic and independent examination to determine whether quality activities and results comply with planned arrangements and whether these arrangements are implemented effectively and are suitable to achieve objectives.

These definitions establish that the essential difference between a review and an audit is that the former is carried out by those responsible for managing the operations being examined whereas the latter is undertaken by some independent body.

The purpose of a **system review** is to carry out an examination to determine whether a system is succeeding in achieving its objectives with a view to changing it if it is failing to do so. It follows necessarily that such an examination can be undertaken only by the management responsible for operating the system since only they can judge its objectives and only they have the power to make changes if these are found to be advisable.

Quality system reviews may be seen as analogous to preventive maintenance programmes for mechanical equipment. They need to be undertaken at prescribed regular intervals, by qualified people following pre-planned procedures and check lists. Their purpose is to reveal and correct defects or irregularities before they cause breakdowns or accidents and to lubricate or eliminate points of friction or overheating. Here is a typical check list for a quality system review:

1. Are the objectives and policies stated in the quality manual still valid, or have they become obsolete due to changes in the business environment?
2. Is the organization structure for quality management functioning satisfactorily and is there proper delegation of responsibilities to nominated personnel?
3. Are the procedures identified and described in the manual appropriate to achieve current objectives and policies?
4. Are the procedures being adhered to?
5. If procedures are being ignored or changed without authority, why is this so and what action should be taken?
6. If procedures are being adhered to, are they achieving the desired effects?
7. What changes in the quality system, if any, are required to make it more effective?

Who should undertake system reviews? One option is that the line manager himself should take the lead, perhaps in unison with the quality assurance manager, or at least with his assistance. The alternative is for the quality assurance manager to carry out the enquiries, assemble the facts and make recommendations to the line manager. Whichever arrangement is chosen, the responsibility for taking action lies with the line manager. It is his system and only he can issue instructions to the people who operate it.

In quality jargon, a quality assurance manager's contribution to a quality review would be termed an **audit**. The essential characteristic of a quality audit is that it is undertaken by independent persons or bodies having no direct interest or responsibility in respect of the organization or project being examined. Audits may be internal, such as those contributing to quality reviews, or they may be undertaken by second parties (purchasers) or they may be carried out by third parties acting on behalf of a purchaser. The latter two types of audit are dealt with in more detail in Chapter 7.

Not all internal audits take place as part of a system review. They may be carried out whenever management has a need for factual information on the extent to which its quality objectives are being met. The executive board of a group of companies may, for example, wish to audit the quality systems of subsidiary companies to prevent any one subsidiary from harming the reputations of the remainder. A company managing director may similarly order audits of projects being undertaken by his organization to ensure that buoyant current profit-and-loss statements are not being achieved at the risk of future heavy maintenance costs or warranty claims.

Internal audits may be undertaken by a specialist team within an organization, or by staff on short-term secondment from other work similar to that being audited, or a combination of the two. Whatever the composition of the team, three factors are of vital importance: the auditors must be trained for the task, the terms of reference for the audit must be clearly defined by those to whom the team will report, and the purpose and mode of conduct of the audit must be acceptable to the management of the work being audited. An internal audit is often seen by those subjected to it as an unwelcome form of prying, and there is a risk that they will exert resistance to thwart its objectives. Legitimization of the activities of the audit team by management is an essential condition of success. After all, if something is wrong, it is preferable that the management should find out about it before the customer does.

Both general-purpose (BS 5750) and nuclear (BS 5882) quality system standards stipulate that internal system audits should be carried out on a regular basis. Their requirements may be summarized as follows:

1. Internal audits are necessary to verify compliance with the quality system and to determine its effectiveness.
2. Audits must be planned and documented.
3. Audits must be performed in accordance with written procedures or check lists.

4. Auditors must be independent of any responsibility for the work being audited.
5. The results of the audits must be documented and brought to the attention of the management of the areas of work audited.
6. The responsible managements must determine the actions needed to correct any deficiencies found.
7. Corrective actions must be followed up within an agreed time scale to verify that they have achieved their purposes.

There can be little doubt that a quality system which is not subject to independent audit lacks credibility. In this respect, quality system standards provide a significant benefit in that they provide both parties to an audit with a common framework within which to work and a standard against which judgments can be made. The practice of quality auditing has become increasingly widespread in recent years as industry's traditional dependence on the inspection function has reduced. Some quality managers spend the bulk of their time either auditing others or being audited. Techniques of both auditors and auditees have become more refined as each tries to gain the upper hand. There are dangers in this specialization. Audits are not an end in themselves and they should not be permitted to become rituals in which the protagonists become locked in battle on the semantics of quality system standards and procedure documents while the real world outside slips into bad habits. The techniques of auditing form the subject of Chapter 9.

Training

No matter how carefully devised and comprehensive a quality system may be, it can only be put into effect by people. If the people do not know how to operate it, or if they do not want to operate it, then failure will result. To give people the knowledge and skill to operate the system, they need to be trained. To give them the will, they need to be motivated.

The construction industry has suffered for many years from a chronic neglect of training. In many ways this is not surprising. The cyclic nature of the construction work load and the nomadic life-style of many construction workers have contributed to an attitude of mind which finds it difficult to look beyond the end of the current contract and which recoils from long-term investment in people. In times gone by, this did not matter very much. Structures were simple and repetitive, and designed with factors of safety large enough to tolerate wide variations in the standards of materials and workmanship. Work which did require particular skills was performed by craftsmen who had learned their trades through long periods of apprenticeship. Contractors maintained their standards through the activities of a core of experienced general and trades foremen, many of whom would have received little formal education but who expected, and merited, the respect of everyone on the site.

Those days have gone forever. Structures are now more complex, design techniques are more precise, factors of safety have been reduced and new materials are rendering the old craft skills redundant. At the same time, more and more work is sub-contracted and intense competition has led to greater speeds of work with less supervision. It is hardly surprising that there has been a general lowering of standards, that some buildings are requiring extensive remedial works within a year or two of completion, and that major structures are facing demolition after lives of only twenty years because they have become dangerous or uninhabitable.

It would be unreasonable to attribute all these woes to lack of training. Equally, there can be no denying that improved skills at every level would make a substantial difference. Whereas, once upon a time, workers such as treadmill operatives or galley slaves might have been able to perform their work satisfactorily solely by the use of muscle power, those days are fortunately past and it is difficult to envisage any task in modern industrial society which cannot be done better by an appropriately trained person rather than by one without such training. In the industrially developed countries of the world there is a clear and direct relationship between national prosperity and training effort.

So, how can our quality system help an organization make more money by employing better-trained people? First, it can require that formalized systems be set up to identify the training needs of each activity and to ensure that people are not allocated to tasks for which they have neither the training nor the experience. The concept that inexperienced people should be 'dropped in at the deep end' of a job in the hope that, by chance, they will learn enough to survive, is archaic and wasteful. Secondly, people can be trained to manage quality. They can be taught the techniques of fault analysis. They can be instructed on how to identify the causes of failure and how to eliminate them. They can be encouraged to assess the costs of poor quality and to make rational decisions on the value of preventive measures. They can be trained to control quality in the same way as they are trained to control costs.

But all of this activity will be fruitless without constant direction and stimulation from the top. There has to be clear evidence that senior management wants work to be done properly and that employees who would seek a quick profit by deceiving a customer are offending against the corporate ethos and will not be tolerated. This cannot be achieved by mere exhortation; words are not enough. People are motivated by what management does, not by what it says.

5

DOCUMENTATION

Purpose

A quality system cannot function effectively unless everybody in the organization knows what it is. Not only do people need to know what they are expected to do, they also need to know what their colleagues are doing in respect of operations which border or impinge upon their own responsibilities. Therefore, to make sure that everyone has a common understanding, it needs to be documented. Documenting a quality system renders it amenable to management control. Companies which depend on unwritten systems for the assurance of quality, and many still do, can find it difficult to cope with change. They rely on long-serving members of staff who know how things have always been done to pass on their knowledge to newcomers. In times when change was slow and when people were accustomed to staying in one job with one employer for long periods, such arrangements could be tolerably successful. This, unfortunately, is no longer true. In many industries, and particularly in the construction industry, staff come and go with great rapidity. There is no time for them to absorb company procedures by word-of-mouth, even if such methods of communication could be relied upon. Furthermore, customer demands, designs, materials and methods are all evolving and changing at an unprecedented rate. Procedures for quality management have to be adapted to meet these new demands. It is the role of management to decide what changes should be made and to inform those people who will be affected. This is the main purpose of quality system documentation. It provides a powerful tool for controlling change.

Definition

The following definition is from BS 5882:

Documentation Any recorded or pictorial information describing, defining, specifying, reporting or certifying activities, requirements, procedures or results.

This definition is in two parts. First it establishes that documentation is 'recorded or pictorial information', and it then goes on to establish the purposes for which such information may be used. Let us take these two aspects in turn.

The essential characteristic of information which is 'recorded' or 'pictorial' is that it has a degree of permanence which enables it to be referred to, studied, analysed, or otherwise made use of, repeatedly and for an unlimited period of time. It ceases to have the potential for error of the spoken word or the human memory. On the other hand, because of its permanence and the possible multiplicity of uses and users, great care must be taken to eliminate that which is false or misleading. The traditional means of achieving permanence and a wide distribution is to commit facts to paper. Drawings, specifications, reports and memoranda have been commonplace in industry since Noah first noted down the specification for the Ark. Nowadays, however, an increasing proportion of documentation is in the form of magnetic tape, microfilm or floppy disk.

Turning now to the purposes of documentation, our definition encompasses two distinct categories of document. Firstly there are those which are produced in advance of the work to which they refer. In the words of the definition, they 'describe, define or specify' and their purpose is to influence or control activities which will take place in the future. In contrast, the second category of documents 'reports or certifies' events which have already happened. Both categories are essential to good quality management, the former to ensure quality control and the latter to provide quality assurance. In practice, however, the divisions between the two categories are not as distinct as might at first sight appear. Each category has a dual role. For example, the existence of a comprehensive series of control documents in itself provides evidence of an effective quality system. Likewise, reporting documents often contain information which will subsequently be used as an instrument of control or to provide feedback which can be used to modify or refine the control documents.

In quality jargon, documents whose principal purpose is to describe or instruct are known as 'manuals', 'instructions' or 'procedures'. Those which report or certify are known as 'records'. The ensuing paragraphs will discuss each of these forms of documentation in turn.

Quality manuals

On p. 49 the purpose of a quality manual was defined as 'to provide an adequate description of the quality management system while serving as a permanent reference in the implementation and maintenance of that system'. To the extent that each company is unique, it follows that every company's quality manual should be an original document. Nevertheless, a pattern has evolved for quality manuals, and a typical example for a construction company is included as Appendix B. The author is aware that he is offering a hostage to fortune in putting forward such a model. Opinions differ on the optimum layout and contents of quality manuals. However it would not be possible to deal adequately with the subject without an illustration of this kind. It is offered as just one example of a document which is both useful and practical and which can satisfy quality system standards. There are many other ways of achieving the same objectives.

The company to which the manual relates (Alias Construction Ltd) is, as its name implies, a figment of the imagination. It is a subsidiary of a large group (The Alias Group) which has widely diversified interests within the construction industry. Alias Construction undertakes medium- to large-size contracts for civil and building works, mainly in the United Kingdom. The Alias Group has introduced a quality policy with which all its subsidiary companies are required to comply and has set out the ground rules for a common approach to its implementation. These ground rules require that manuals should be succinct, easy to read and as brief as possible. The amount of detail is to be kept to a minimum in order to emphasize important matters and to avoid the need for frequent revisions. In keeping with the Group's corporate image, manuals and quality plans are required to display a uniformity of appearance. They therefore follow an established format and are typed on a pre-printed sheet which displays the Group's logo and has standard locations for the document heading, document number, page number and date.

Each section of the manual will be discussed in detail in subsequent paragraphs, but before embarking upon this it is as well to clarify the uses for which it is designed. These are two-fold:

1. To inform staff within the organization of the quality policy which has been adopted by management and to advise them of the means by which the policy will be achieved.
2. To demonstrate to clients and purchasers that the organization operates a quality system capable of assuring the quality of its products or services.

The manual is in five sections:

1. Control
2. Corporate philosophy
3. Company organization
4. Company standing instructions
5. Project quality assurance.

It is issued in a good-quality ring binder which complies with the Group's corporate image and displays the company's logo.

TITLE PAGE

The title page serves a number of functions. It identifies the company's name and address, it quotes the document's code number (which is also its word processor file reference) and it indicates its issue status. It also has spaces for the name of the person to whom the manual is issued and the unique copy number allocated to him. Note, too, that the title page also contains two statements, one which concerns copyright and the other which identifies the contractual and legal status of the Manual. The purpose and background of these statements are discussed further on p. 85 under the heading 'Contractual and Legal Aspects'.

Immediately after the title page there comes a signed statement by the Chairman and Chief Executive. The purpose of this statement is to substantiate the document's authority and to demonstrate management commitment at the highest level to achieving the objectives which it sets out. It makes clear, at the outset, that the quality system is to be taken seriously and that all employees are required to play their parts in making it effective.

Next comes the Table of Contents. This serves the conventional function of providing the reader with a guide and reference to the structure and content of the document.

1. CONTROL

This section introduces the manual to the reader, establishes its authority and describes the method adopted to keep the manual up-to-date and to inform users of changes that have been made. Quality manuals are living documents, and just as an organization has constantly to adapt itself to changing circumstances, so too must the manual which describes its quality system. However, while changes are necessary, they can also present a hazard unless they are controlled. Methods of change control for documents are discussed further on p. 79.

2. CORPORATE POLICY

This section is an extract from the Alias Group Quality Manual and it is incorporated in all Alias company quality manuals. It states the corporate objectives and establishes the principles with which company quality systems are expected to comply. It then goes on to describe the arrangements made at Group level to monitor and report on compliance with quality policy.

3. COMPANY ORGANIZATION

This section describes the activities of Alias Construction Ltd and sets out its management structure. It defines the responsibilities attached to the principal managerial posts and establishes the status and duties of the company quality assurance manager.

4. COMPANY STANDING INSTRUCTIONS

This section schedules and describes all company standing instructions which relate to the management of quality and cross-references them to the relevant paragraphs of BS 5750:Part 1 and BS 5882. The purpose of the cross-reference is to facilitate the task of external auditors who will wish to satisfy themselves that each of the requirements of the standard against which they are auditing has been addressed in the company's documentation. As an alternative to providing a cross-reference, some organizations produce a procedure covering each of the clauses of the standard they are seeking to satisfy. While this may be of considerable benefit to an auditor, it is likely to detract from the value of the procedures as instruments of management. It also ties the documentation to one particular standard.

Note that of all the standing instructions scheduled, only the five issued by the Company Quality Assurance Manager could be described as 'quality assurance procedures'. All the rest are instructions by management as to how work should be done. Alias Construction has avoided the trap of producing 'QA procedures' whose only purpose is to satisfy a quality standard. Such procedures encourage the misconception that quality assurance consists of an additional bureaucratic routine designed to satisfy auditors. As the schedule shows, the Alias Construction standing instructions do in fact address all the requirements of the two standards referred to. This, however, was a secondary objective in their preparation. The first was to improve the performance of the company.

Alias Construction favours the term 'company standing instruction' in preference to 'company procedure' mainly because it already had a set of standing instructions in operation within the company at the time that the decision was made to formalize its quality system. The newly-appointed company quality assurance manager discovered that many of the topics which needed to be addressed were already dealt with quite satisfactorily in existing standing instructions. He therefore proposed that rather than create a new set of procedures serving the quality system only, he would graft the additional documentation on to the stem of the existing system. His proposals were accepted and, because of his obvious talent, he was invited to take on the harmonization and administration of all company standing instructions.

5. PROJECT QUALITY ASSURANCE

This section sets out the procedure for the assignment of quality assurance engineers to projects and outlines their duties. It then describes the preparation of project quality plans and project procedures.

Quality plans

The purpose of a quality plan is to specify how the quality system described in the quality manual will be applied to a particular project and to give details of the specific practices, resources and activities which either have been or will be developed for this purpose. Quality plans for design work and for construction are discussed further in Chapters 6 and 8 respectively. A specimen quality plan for a small construction site is attached as Appendix D.

Standing instructions and procedures

It is one of the most essential of management tasks to ensure that those responsible for executing work should know what they have to do and how they should go about doing it. Quality system standards require that this should be achieved by the 'establishment, maintenance and implementation' of 'documented procedures and instructions'. The use of both words 'procedure' and 'instruction' implies that there is a difference in meaning between the two, but this is not explained in the current version of BS 4778. The following definitions are from the 1979 edition of BS 4778 and from BS 5882 respectively.

Instruction The written and/or spoken direction given in regard to what is to be done, including the information given in training.

Procedure A document that specifies or describes how an activity is to be performed.

These definitions identify two distinctions. Firstly, an instruction decrees *what* is to be done but a procedure tells *how*. Secondly, a procedure is always documented but an instruction need not be. The latter distinction is perhaps petty, but the former is important since it indicates that whereas all instructions are by definition mandatory, a procedure need not be so unless accompanied by an instruction which says 'do it this way'. In practice the words are virtually interchangeable, and it would be pedantic to dwell further on the matter. It is interesting to note, however, that the Alias Construction Ltd Quality Manual contrives to make all procedures mandatory by including the following words in para 1.2 'Authority':

> 'All personnel ... shall perform their duties ... in compliance with company standing instructions and project procedures.'

In most companies, instructions fall into two categories: standing instructions or procedures, and work instructions. Instructions in the first category prescribe the routines governing departmental or inter-departmental operations which normally remain constant and are applied to all projects being executed or all products being manufactured. They might cover, for example, an organization's procurement system, the rules for hiring and firing staff, the procedure for approving and paying personal expenses claims, or the arrangements for checking design work. Some, such as the first and last of these examples, can have a direct influence on product quality, and these should be included, or referred to, in the company quality manual. Work instructions, as the name implies, relate to specific work elements or products. They are discussed in more detail later in this chapter.

The purpose of standing instructions and procedures is to ensure uniformity of understanding and performance and to provide continuity when personnel changes occur. They define what is to be done, how it should be done, who should do it, and when. Since they are intended to be mandatory, provision has to be made for enforcement. It follows that they must be issued under the authority of the management responsible for the work which they cover. This will normally be the relevant departmental or project manager. Standing instructions which cross departmental boundaries are best prepared by the appropriate functional head and endorsed by the chief executive.

It is an unfortunate fact that many excellent managers and supervisors baulk at the task of producing written instructions for work for which they are responsible. They know what has to be done and can tell people what to do, but they lack the ability to organize their thoughts on paper. Quality managers are often pressed to prepare instructions for line managers who plead lack of time or resources to produce them themselves. Such pressures should be resisted. A quality manager who produces and issues instructions in respect of work for which he is not responsible is not only usurping the role of the manager who should be responsible, he is also collaborating in an activity which will inevitably weaken the work-force's support for the system. It is a pre-requisite of success that production management should support the aims of the quality system and communicate this support to their subordinates. A production manager who, however unwittingly, portrays the attitude that procedures are the prerogative of some other person who is trying to impose additional burdens will soon find his work-force devising ways to frustrate the system rather than co-operating in making it work.

There is a further benefit which can accrue to line management through the production of documented instructions. Organizations which are accustomed to operating through custom and practice communicated by word of mouth frequently have a propensity to indulge inefficient practices which are followed unquestioningly because they are part of a tradition. Very often the assembly of facts, the thought and the analysis necessary to prepare a written instruction can expose areas of weakness or duplication which, if changed, can result in significant savings of money through the reduction of waste, repairs and delays.

Most construction work is carried out on a project basis. Thus, contractors can expect at any one time to be working on behalf of a number of clients and on a variety of different classes of work. Each project may be governed by its own particular form of contract and clients also have the right to impose their own preferred methods of operation. Circumstances inevitably arise when the contractor's standard procedures are clearly unsuitable or inadequate for application on a particular project. There is therefore a need for a mechanism which can provide for the preparation and issue of project procedures, either by amendment of company procedures or by the creation of new procedures in respect of subject matter for which no company procedures exist. This subject is addressed under the heading 'Project Quality Assurance' in the company quality manual of the fictional Alias Construction Company which is attached as Appendix B.

To make the task of preparing procedures and instructions less daunting to the busy manager, and to achieve a degree of uniformity in their format, Alias Construction also issues a 'procedure for procedures'. This is identified in the table on page 13 of their quality manual as CSI QA-02 'Preparation and Administration of Instructions and Procedures' and a copy is attached as Appendix C. As with the quality manual itself, the format it defines is that which has been found acceptable in other companies within the Alias Group and this procedure is included in the documentation of all Alias companies with only minor modifications to suit particular markets. Again, it is offered to the reader as just one way of preparing documents which can function both as useful tools of management and as a means of satisfying the requirements of quality system standards.

Appendix C is designed to be self-explanatory and its contents will not be discussed in detail. However, it is worth summarizing the nomenclature which Alias Construction has selected for the different categories of instructions and procedures, and to note the alternative and equally acceptable terms commonly in use for the same documents:

Company standing instructions These give directives on standard departmental or inter-departmental methods of operation for head office and project activities which remain relatively constant regardless of the type of work currently being done.

Alternative terms are:

Company standard procedures
Corporate procedures
General procedures.

Project procedures These specify standard methods of operation to be used on a particular project or site. Such methods will normally follow standing instructions but occasions inevitably arise when, for one reason or another, these are not suitable. In such cases, project procedures are created either by amending standing instructions or by preparing new documents.

Alternative terms are:

Project standing instructions
Site procedures
Site instructions.

Work instructions

The instructions and procedures discussed in the previous pages related mainly to procedural and organizational matters which remain reasonably constant even in a construction organization undertaking a variety of different classes of work. We now consider a further tier of documents whose purpose is to ensure that those responsible for executing work understand the specification requirements and receive clear instructions as to how they should go about their business.

There is a number of ways in which such work instructions can be conveyed. Some organizations depend upon the spoken word. For example, when Trooping the Colour, the Guards Regiments rely solely on verbal commands to order a large body of men to perform complex manoeuvres with precision and style. But there are limits to the effectiveness of the verbal method. The number of different instructions needed to troop a colour without serious mishap is quite small, probably not more than thirty, and the order in which they will be given can reasonably easily be anticipated. Even so, it is necessary for the personnel concerned to undergo a lengthy period of training and rehearsal before they can perform to specification and satisfy the customer. This is not to decry the parade ground method: there will always be a need for direct personal command, but most organizations, including the army, realize that there is no substitute for the written word for communicating any but the simplest of instructions.

Here are some examples of documented work instructions which may be found on a typical construction site:

Drawings issued for construction
Contract specifications
Construction method statements
Temporary works drawings
Wall posters
Welding instructions
Maintenance manuals
Sketches from an engineer's duplicate book.

This is an extensive list, and it effectively counters the argument that paperwork introduced as part of a quality system adds unnecessary burdens to the construction site. The paperwork, and the bureaucracy, are with us already. Some is essential, some less so, some could no doubt be dispensed with altogether. Clearly there is a need here for control. The question is: How much? Some guidance is offered by quality system standards. BS 5750 requires that processes which affect quality shall be carried out under 'controlled conditions'. It then goes on to say that controlled conditions include:

> 'documented work instructions defining the manner of production and installation, where the absence of such instructions would adversely affect quality'.

The above extract indicates an acceptance of the need for judgement in deciding whether a documented work instruction is necessary for the control of a particular task. The standard does not say that every activity on the site has to be covered by a written work instruction. There can be no dispute that the absence of drawings, or specifications, or welding instructions would have an adverse affect on quality. But to stretch this assumption to argue a need for a written instruction saying, for example, 'Remove the rubbish from that hole and put it into the skip' would defy common sense and would not be a reasonable interpretation of the standard's requirements.

So, what steps should be taken to make sure that work is adequately covered by documented instructions? Here are some suggestions:

1. Examine the drawings and specifications. Are they complete, and do they show adequately what is to be done and how? If not, additional instructions and drawings must be made.
2. Is the work of a kind that can be performed satisfactorily using current good practice and the craft skills likely to be available? If not, then special instructions must be issued.
3. Is any of the work of an innovative nature or does it use conventional materials in an unconventional manner? If so, people must be made aware of any new techniques which are required.
4. Did the planning and pricing of the work require the development of special construction methods to cope with specific problems? If so, these should be committed to paper and issued to the Project team.

On a well managed project, the above matters will be attended to at the planning stage. After deciding what is to be done, the planning staff should prepare and issue properly drawn sketches or flow charts, giving clear and complete information to the recipients. The list of examples of typical instructions on p. 78 ends with 'sketches from an engineer's duplicate book'. On a well-planned site, such instructions should not be necessary, still less should there be instructions written on the backs of envelopes or on cigarette packets. Instructions of this kind are not 'controlled conditions' as required by the standard, nor do they have any place in an effective quality system.

Document control

All the documents mentioned in this chapter are subject to change. If these changes are not controlled, the system will break down. This should not be interpreted as a disadvantage of a documented system. Change is inevitable, and it is easier to revise and re-issue written instructions than it is to countermand a verbal order.

Document control commences at the point of origin. This may be the drawing or planning office for pictorial documents, or it may be the office of a departmental head responsible for the issue of standing instructions. There is a need for a documented procedure which identifies the persons responsible for the original preparation and approval of documents and which ensures that subsequent changes are reviewed and approved, either by the originators themselves, or by other persons who are both properly authorized and in possession of all the relevant background information. The procedure should go on to establish how changes are to be recorded, how all recipients should be informed, how obsolete documents will be removed from circulation and what records must be kept.

Procedures for controlling changes to drawings or other single-page documents are familiar in the construction industry and are dealt with briefly on p. 101. The control of multi-page documents such as manuals or procedures is a more complex problem. Changes may be minor, affecting only one or two words on a single page, or they may involve major rewriting or rearrangement. Many organizations operate systems whereby amendments are dealt with by the publication of revised versions of the appropriate pages followed by a general re-issue only when the number of revisions becomes excessive. These systems rely on the diligence of the recipients for the correct insertion of revised pages and the discard of those which are obsolete. When establishing its systems, Alias Construction decided that since its quality manual and standing instructions were fairly slim documents and were not expected to be subject to rapid change, it would be more cost-effective to re-issue these documents as a whole when this became necessary rather than attempt to revise them page by page. The system they decided to adopt is described in Section 4.7. of the standing instruction enclosed as Appendix C.

Organizations in receipt of documents, such as construction sites, should establish similar procedures for controlling the receipt and re-issue of documents. On a small site, a conventional drawing register is usually all that is needed. Large sites, particularly those employing substantial numbers of sub-contractors, may find it beneficial to introduce computerized systems.

The ease with which documents can now be duplicated can render control more difficult. The recipient of a controlled document who prints and issues uncontrolled copies makes it impossible for the appointed register holder to keep track of what is happening. Similarly the user of an uncontrolled copy may be innocently unaware that the version of a document in his hands has been superseded. There is thus a need for strict discipline to be imposed on the recipients of controlled documents, with severe penalties being imposed on those who make unauthorized copies.

Not all copies of controlled documents need to be controlled. For example, copies of quality manuals may be issued to clients as part of a prequalification submission. It would be neither practicable nor necessary to trace such copies or to follow up with amended versions. In such cases, the normal practice is to stamp copies with the words 'uncontrolled copy', or 'not subject to change control'. No record of issues need then be kept and users will be aware that the document in their possession may no longer be valid.

Records

The purpose of compiling records is to chronicle what has taken place and provide reliable, factual information for future use. In a quality system, records are required in order to demonstrate that the system is functioning satisfactorily and that the required standards are being achieved. They supply objective evidence to enable a manufacturer or contractor to exert control over his operations and they provide verification to a purchaser that the goods or services he is buying will comply with his requirements.

Records supply alarm signals to warn of dangers or adverse trends. They supply the factual information needed for statistical quality control. They provide input for cost control. They provide proof of compliance with specification. Whether presented on paper or on more exotic media such as microfilm, magnetic tape or floppy disk, they provide the facts and information without which no system can operate. Consider two categories of record:

1. Records whose primary purpose is to control future actions.
2. Records whose primary function is to provide factual information.

It is necessary to insert the word 'primary' into both these category descriptions since many records fulfil both functions to a greater or lesser extent. Take, for example, a documentary record used in a service industry: a travel ticket. Its primary purpose is to control access to a means of transport, be it train, flight or coach. It provides evidence that the holder has paid the requisite fare and enables staff to divide those entitled to travel from those not so entitled. However, once the journey is over, it has served its purpose and will be discarded. Its value as a control document is vital at the time, but its use as evidence is short lived because the benefit it bestows is ephemeral and the information it contains of little interest to posterity.

Compare the latter example with the records kept of the construction details of a multi-storey commercial building. The life of such a structure may be a hundred years or more and it is likely to undergo many changes of use during this period. Suppose it is decided to install a communications dish aerial on the roof, or an air-conditioning unit. The original designers could not possibly have contemplated such an eventuality and the structure may be quite unable to support the additional loads without being strengthened. In such cases, as-built records are invaluable to the user of the building. Although not compiled with specific control purposes in mind, they are worth all the effort of preparation and storage, even though those who prepared them could not have imagined the eventual use to which they would be put.

These two examples relate the purpose of a record and the type of information it contains to the period for which it should be retained. All too often, for lack of careful thought, records are either discarded prematurely or stored well beyond the expiry of their practical value. In the case of quality records, the latter seems more often to be the case and this is perhaps the reason why, in the minds of many people, the operation of a quality system has become associated with a proliferation of unnecessary records. Such views are often encountered among those who have experienced the operation of quality systems imposed upon suppliers by purchasers. All too frequently, such mandated systems become unreasonably bureaucratic as purchasers seek to eliminate every conceivable possibility of non-conformance and to require paper verification of every requirement. There is a need for balance and a rational approach towards the generation and subsequent storage of records.

Clients, consultants and contractors are subject to varying influences which determine their policies in respect of the maintenance of records. Contractors keep records to substantiate requests for additional payments or extensions of time and to guard against claims for compensation by clients or others arising from alleged defective work. Consultants, too, require records for defensive purposes, and they need data to enable them to deal justly with contractors' claims. They also have a need for records which preserve knowledge of designs and techniques which may be applied on future projects.

Clients need records which will correctly identify the 'as-built' condition of the structure for retention during its lifetime. If specific information is required for operational or maintenance purposes, this should be defined in advance by the client or his advisers so that the appropriate manuals can be compiled as work proceeds. Records may also be necessary in order to gain the approval of regulatory authorities. In the case of nuclear facilities, it is a requirement that records 'shall be maintained by or for the owner for the useful life of the item from manufacture through storage, installation, operation and decommissioning', and that 'records shall be stored and maintained in such a way that they are readily retrievable in facilities that provide a suitable environment to minimize deterioration or damage and to prevent loss' (BS 5882).

In non-nuclear work, commercial and legal considerations normally prevail in determining the types of record to be kept and the times of retention. Page 32 outlines the periods of limitation during which a plaintiff may pursue claims for negligence causing latent damage. These have a 'long stop' maximum of fifteen years and this is the period most frequently used to govern retention times for records. However, the 'long stop' provision only governs the period within which proceedings must be issued. In the case of High Court proceedings, the defendant need not become aware that proceedings have been issued against him until a writ is served, and the rules of the Supreme Court permit service of a writ at any time up to one year from the time of issue. To allow for this eventuality, it is wise to add one year to the fifteen years period when deciding on the duration of protection to be adopted.

A further trap for the unwary is posed by the possibility of a claim under the Civil Liability (Contribution) Act, 1978. This stipulates that a defendant, against whom judgment has been awarded, may recover a contribution towards his damages from a third party who is wholly or partly to blame for the plaintiff's loss. Under the Act a defendant may commence a contribution action at any time up to two years after judgment has been made against him. Bearing in mind that it may take several years for the original action to come to trial, consultants and contractors may be in jeopardy as potential third parties for very much longer periods than those laid down in the Latent Damage Act 1986.

Let us consider first the records maintained by contractors' agents and resident engineers to support or adjudicate on claims for additional payments or extensions of time. Procedures for dealing with such claims are outlined in standard conditions of contract. Records kept for claims purposes include:

Daily diaries
Start and finish dates of each activity
Allocations of staff, labour, plant and other resources
Normal hours worked and resources working abnormal hours
Weather affecting the work
Strikes or other industrial disputes
Causes and extent of delays
Investigations, tests and inspections
Consents and permits
Dates of dispatch and receipt of drawings
Issue and processing of requests for variations.

These documents may be discarded once final payment is made.

A second category of records are those which may be required to substantiate a case in court. They include:

Original contract documents
Sub-contracts
Sub-contractor warranties
Orders on suppliers
Site meeting minutes
Site manager's diary
Resident Engineer's diary
Correspondence between Contractor and Architect/Engineer
Correspondence between Client and Architect/Engineer

Correspondence between Client and Contractor
Sectional and final Certificates of Practical Completion
Certificates of making good defects
Documents confirming concessionary acceptances
Final issues of construction drawings
Final Certificates of Completion.

All the above should be retained for at least sixteen years by the contractor or consultant as the case may be, and should then be destroyed only after consideration of any potential litigation which might require their subsequent production in court.

Finally, let us consider the records generated during the course of the work to verify compliance with specification. These will include:

Materials conformance certificates
Inspection reports
Laboratory test results
Piling records
Concrete placing records.

These all contribute to the issue by the Architect or Engineer of a final completion or maintenance certificate at the end of the contractual maintenance period. Once this has been issued, it effectively supersedes the quality control records, which thereby become redundant and could, in theory at least, be destroyed. Whether this would be a wise move in any particular circumstance must depend on the contractual arrangements. Some standard conditions, such as the JCT Standard Form, state that subject to some qualifications a Final Certificate may be held to provide conclusive evidence that the quality of materials and standard of workmanship are to the satisfaction of the Architect or Supervising Officer. The ICE Conditions of Contract, however, state that the issue by the Engineer of a Maintenance Certificate should not be taken as relieving the Contractor or the Employer of any of their obligations to each other. The possibility of a later need to produce quality records in the event of litigation should therefore be considered before any are destroyed. On the other hand, the sheer volume of space occupied by quality records and the consequent storage costs are sufficient to persuade many contractors and consultants to dispose of them at the earliest opportunity.

The archive storage of documentation should be considered at the start of a project and filing systems established to select material for long-term storage as it is generated. If this is done, all that is needed at the end of the contract is to compile a register scheduling the content and location of each file. Storage facilities should be secure, dry, fire-proof and protected from vermin. To reduce the volume of archive records, documents may be microfilmed. However, in order to be admissible as evidence in legal proceedings, the production of microfilm must be shown to be part of a normal business procedure. To do this, the following criteria should be observed:

1. Documents microfilmed must be in a logical order without any omissions.
2. An authorization certificate (signed by a responsible official and the camera operator) must be included at the start and end of each run of film. This certificate should state that the microfilms are being made as a normal business practice.
3. Once filming has been completed, the microfilm must be checked to ensure that it is an accurate record of the documentation. A written record of these checks must be maintained and signed by the checker.
4. On completion of the above, the responsible manager may give authorization for the original documents to be destroyed. A written record of what has been destroyed should be maintained.

It is best to keep two copies of each microfilm in separate locations and in fire-proof cabinets. If one set is 'jacketed' for ease of reference, the other should be kept uncut as the master copy.

Contractual and legal aspects

One of the purposes of a quality manual and related documentation is to provide information on an organization's quality system for the benefit of potential purchasers. Manuals may be offered as part of a sales package, or they may be given in response to a call for pre-qualification information, or they may be submitted with a tender. What is the contractual status of such documents?

On p. 31, reference was made to the power of the courts to impose 'implied terms' of contract. A document or statement issued or made prior to the signing of a contract can be held to be a 'representation' and enforceable as an implied term of the contract. It is conceivable that quality manuals and similar documents could be viewed in this light. A purchaser might then be able to argue that a contractor had an obligation to implement procedures contained or listed in a quality manual, and that departure from such procedures without approval would be a breach of contract. This would tilt the contractual balance steeply in the purchaser's favour and limit the contractor's freedom to select and develop the most economic methods of achieving specified standards of materials and workmanship. To guard against this eventuality, it is advisable to clarify the status of quality manuals and other documents by stating on the title page that they do not form part of any contract and are not intended to imply any representation or warranty. The quality manual of Alias Construction (Appendix B) and the specimen procedure (Appendix C) both have such a statement. They also have a statement which establishes ownership of the documents and their contents and prohibits disclosure without the consent of the company.

Quality managers would do well to consult their company legal advisers on these matters, since the law is open to different interpretations and its application in specific cases can be difficult to foresee. Quality system procedures are not usually written as legal documents, indeed they would probably be useless if they were. It is therefore wise to clarify their status at the outset so that the chances of later misunderstandings are minimized.

Moving from the general case to the particular, there are circumstances when the instructions and procedures described in a quality manual are required to be contractually enforceable. This applies particularly in the case of nuclear work where contractors may be required to submit outline descriptions of their quality systems with their tenders and be prepared to agree contractually binding arrangements if successful. Obviously, in such cases, the recommended disclaimers would not be appropriate and should be removed once negotiations are at an end and purchaser's approval has been granted.

The application of quality assurance to mainstream construction work is still in its infancy and possible conflicts arising from the invocation of quality system standards alongside standard conditions of contract have yet to be tested in the courts. Meanwhile it is wise to take reasonable precautions.

6

DESIGN

The design process

A design organization may be likened to a factory. Raw material in the form of design briefs, standards and codes of practice are fed in at one end. These are then processed by designers and computers and an output is produced, consisting mainly of working drawings and specifications. This analogy is perhaps controversial and some may question its total validity. However, it is sufficiently near to the truth to illustrate that the principles of good quality management are as relevant to the design of building and civil engineering works as they are to their construction. Furthermore, the analogy serves to focus the mind on the chain of events which must be controlled if a design office is to meet the requirements of a purchaser. But, who is the purchaser?

An immediate answer to this question could be that the purchaser is the person or organization acting as client, to whose brief the work is done and from whom payment will be expected. Such a response is only partially correct. An architect commissioned in the conventional way may be paid by his client, but the recipient of his product is the building contractor engaged to construct the works. An architect who concentrates on satisfying the aesthetic, economic and functional requirements of his client, while neglecting to provide clear and unambiguous documents to the builder is not doing his job properly. He is equally at fault if he fails to produce work on time to meet the construction programme, or if his designs do not take account of the practical limits of the construction techniques and skills likely to be available. This is an argument in favour of the combined design-and-construction contract in which the designer is hired and paid by the contractor to receive the client's brief and to prepare drawings and specifications for the contractor. Such an arrangement clarifies the designer/contractor interface and renders it easier to control.

Design quality systems

Design organizations may stand alone, for example as firms of architects or consulting engineers, or they may be departments within companies for whom design may be only one of a number of activities. An example of the latter would be an internal design department of a housebuilder or contractor. Such departments may carry out the design of permanent structures or temporary works, or both. Quality management is as germane to temporary works design as it is to permanent works.

Although design control is addressed in quality system standards as just one of a number of activities which should be covered in an organization's quality system, it is recommended that in most cases it is preferable to treat a design office as an organization in its own right, even if it does not exist as an independent body. This is the arrangement adopted by the design department of the fictitious Alias Group, the parent company of Alias Construction Ltd whose quality manual is enclosed as Appendix B. Alias Design, which is a division of Alias Services Ltd, has its own quality manual to describe its quality system. The manual is similar in format and content to Appendix B and it includes the same Section 2 'Corporate Policy' which describes the quality philosophy of the Alias Group. Note that the Alias Construction company standing instruction entitled 'Design of Permanent Works' (CSI OPS-07) is concerned solely with the commissioning and control of design consultants engaged by the company (including Alias Design) and does not deal with matters of design control. CSI OPS-08 'Design and Construction of Temporary Works' does include some design control procedures, but these relate only to the site design of minor structures.

A quality manual of a design department should contain a statement of authority signed by the Senior Partner or whoever is in charge. It should illustrate and describe the organizational structure, giving brief job descriptions of the principal managerial roles. It should include a schedule of standing instructions or procedures together with brief descriptions of their scope. To ease the task of auditors, the schedule could be cross-referenced to the requirements of quality system standards. The delegation of responsibility for quality assurance in a design organization requires careful thought. All design offices have to contain their overheads and only the larger consultancies can be expected to sustain full-time quality managers. The arrangements adopted will depend on the type of work undertaken and on the verification and documentation requirements of clients, statutory bodies and certifying authorities. The latter are particularly onerous in nuclear and offshore work. The following guidelines are suggested:

Offices of less than 20 persons

In a small office, the person in charge has to take personal responsibility for many of the functions which, in a larger organization, would be delegated to others. He is likely to do his own marketing, his own administration and his own technical supervision. He himself will perform such calculation checks as are necessary and will sign off the drawings. In these circumstances he will be operating his own quality system and will be his own quality manager.

Offices of 20 to 100 persons

As an organization grows, there comes a point at which it ceases to be practicable for the person in charge to check everything done on his behalf. He then needs to establish a quality system and to delegate its operation to one of his staff. This need not be a full-time appointment. In a small partnership, for example, the role may be combined with that of librarian, or research officer. In the case of a contractor's design office, one quality assurance manager may serve two or three departments or subsidiary companies. The essential criteria are that he should not have executive responsibility for budget or programme, that he should have the confidence of the management and that he should be able to win the respect of those upon whose co-operation he will depend.

Offices of more than 100 persons

Large organizations require a full-time quality manager, although the break point at which this becomes necessary will vary considerably according to the field of practice. The role of a design office's quality manager will be similar to that of a company quality manager as outlined on p. 60.

Administration

A design office's Administration Procedure should describe all the non-technical routines necessary to introduce work into the office, to keep track of what is happening, and to ensure that the bills are paid. Here is a typical table of contents:

1. Introduction
2. Scope
3. Personnel Responsible
    - Administration Manager
    - Commercial Manager
    - Project Managers
    - Chief Draughtsman

4. Procedure
   Collection and distribution of mail
   Facsimile messages
   Telex messages
   Filing
   Cost estimates
   Time-sheets
   Expenses
   Printing
   Leave of absence
5. References
6. Exhibits.

'Exhibits' should include copies of all standard forms, calculation sheets, drawing title blocks, etc. used in the office.

Review and confirmation of the brief

A construction project comes into existence to meet the needs of a purchaser or owner. Conception occurs when purchaser meets designer and explains to him what he requires. At this point purchasers usually know roughly what they want and how much they are prepared to pay. Unfortunately, and this is particularly true of those who only rarely enter the construction market, many purchasers find it difficult to articulate their requirements in a form which will enable the designer to devise an optimum solution. As a result, misunderstandings occur which can have a damaging influence on the quality of the project which cannot be corrected or compensated for at a later stage.

It is therefore necessary for a design office to follow a procedure which will enable the right questions to be asked at the right time to elucidate the information which is required. Such questions may include:

What is the purpose of the project?

One definition of quality is 'fitness for purpose'. If the designer of a structure does not know and understand its purpose, it is most unlikely that his proposals will fit. It may be that the purchaser himself does not know, or has not decided, what the purpose will be. Perhaps he does not intend to use the project himself, but to sell or rent it to someone else. In such circumstances the onus is on the designer to establish the criteria to be satisfied and to obtain the purchaser's agreement to these before proceeding.

What aesthetic qualities are required?

This is a matter of great significance in some designs, but less so in others. For example, many building owners require structures solely to create a suitable internal environment, in which case their primary interest is to maximize volume for minimum cost. Other purchasers may require a style and appearance which will reflect their business image and add to their prestige. Maybe the project will be subject to the approval of an environmental protection body such as the Fine Arts Commission.

What loadings will structures have to support?

Before he can start work, a designer needs to be aware of all the forces which are likely to impinge upon his structure. These will include standard loadings such as those specified by highway departments for bridge structures. Account has also to be taken of external environmental conditions such as wind and snow. Other loadings may have to be deduced from the intended use of the structure. For example, in a factory, what equipment will be installed? What will it weigh? Will it create vibrations? Will it require special air conditioning equipment? The client himself may be unable to respond to these queries and may refer the designer to other sources of information such as equipment manufacturers.

What information is available?

Depending on the stage in the development of the project at which the designer receives his brief, much of the technical data required for his design may already have been obtained. This could include:

Feasibility studies
Topographical surveys
Site investigation reports
Hydrological or hydrographic reports
Traffic surveys
Details of adjoining structures
Details of existing services.

If vital information is not available, it is necessary at this stage to advise the client of the steps which should be taken to put the necessary studies in hand.

What funds are available?

In any design there has to be a balance between perfection and economy. Purchasers like to have both, but this is not possible. The designer has to perform an iterative process to reach the compromise most acceptable to the purchaser, and it helps if he has a target to aim at.

What is the required lifetime of the project?

A designer is entitled to know the projected lifetime of the structure he is designing, and the policy which will be adopted towards its maintenance. Does the purchaser wish to optimize the combined costs of initial construction and lifetime maintenance, or is he more interested in minimum first cost? Such considerations can have a significant impact on design.

These are just some of the questions which a designer may wish to raise. Only rarely does he get all the answers before he starts work. More frequently, the purchaser's requirements become available over a period of time as design meetings are held to discuss specific queries. The decisions reached at such meetings have commercial as well as technical implications. Questions may arise from the first creative conceptual studies and the answers to such questions should be confirmed in writing in accordance with a procedure agreed between the designer and the purchaser.

Having asked the necessary questions and recorded the responses, it is useful to assemble the design basis on a standard format. Figure 6.1 illustrates a specimen check list which may be used for this purpose.

Design planning

Once the brief is established, the next step is to draw up a Quality Plan for the work. This should identify the resources to be deployed and indicate how the work is to be controlled to ensure technical integrity and compliance with programme. A design Quality Plan may include the following:

1. The names of the staff members delegated to manage the work and to take professional responsibility for its adequacy.
2. Summaries of the delegated responsibilities of the staff named above.
3. The name of the client and his nominated representative, if any.
4. A summary of the design requirements identified in the Client's brief. This may comprise a standard form such as Figure 6.1, together with attachments.
5. A bar chart or similar illustration showing phases of the work and key dates for the receipt and provision of data
6. A schedule of the departmental and other procedures to be implemented to ensure technical adequacy. These will include the standards, specifications and codes of practice to be followed, and the regulations of statutory bodies with which compliance will be necessary.
7. The procedure to be used to measure and control the progress of the work.

A

DESIGN INFORMATION

Sheet of

Date

Made By

Checked by

CONTRACT NUMBER

CONTRACT:

Client

Architect

Controlling Authority

Building Regulations and Design Codes

Intended Use of Structure

Location

Fire resistance requirements

Location	Live Load	Partition	Finishes	Other eg. Dynamic
Roof				
Typical Floor				

Applied Loads

Speed

Factors

Wind Loads

Soil report ref.

Subsoil conditions

Design bearing pressure or Pile capacity & type

Foundation type

Location	Below ground	Submerged	Superstructure

Exposure conditions

Concrete	Foundations	Superstructure	Precast

Reinforcement

Other

Material Properties

Figure 6.1 Design information check list.

8. A communications matrix for correspondence, minutes of meetings, drawing and specification control, design change notes, technical queries etc.
9. Arrangements for consulting with the Client and for obtaining approvals when necessary.
10. A schedule of design reviews and system audits.
11. A schedule of records to be produced and their retention periods.

Provision should be made for reviewing and updating the design Quality Plan as work proceeds. It should be a 'controlled document' (see p. 79).

Conceptual design

In contrast to most other aspects of design and construction, the quality of conceptual design is not generally susceptible to improvement by working to predetermined procedures. The creative impulse cannot be programmed in advance.

A client wants a river crossing. Should it be a bridge or a tunnel? If a tunnel, should it be bored or sunk as submerged tubes? If bored, should it be round in section or horseshoe shaped? If round, should it be lined with cast iron or concrete? One could set up a flow chart to guide the designer in arriving at the optimum solution to this sequence of decisions, identifying the criteria to be applied at each point of choice. It is doubtful, however, if such a chart would be of much assistance to an engineer fully experienced in the field of work. Once appraised of all the relevant information, he would arrive at his decision apparently as if by instinct, but in fact it would be the sub-conscious sifting of years of experience of similar problems within his mind which would lead him unerringly to the correct conclusion. Prudence would no doubt then lead him to check his decision by costing the alternatives, but this would come after the creative flight of fancy, not before.

If this be the case with a prosaic example such as that given, still more is it true when decisions have to be made on matters of beauty or shapeliness. An architect is asked to design an office block. He is shown the site and told the area of space required. Should he make it tall and thin or short and squat? Should he clad it in brick or glass? What colours or textures should he select? An architect's reputation is based on his ability to decide upon issues such as these. His decisions will reflect his years of training and experience and his artistic flair. Planning and procedures may assist these attributes but cannot compensate for their lack. It implies no disrespect to the techniques of quality assurance to note that Michelangelo succeeded in painting the Sistine Chapel without the benefit of such concepts as change control, non-conformance reports or design audits.

Design control

Once the basic concepts have been established, the next steps are to carry out such structural and other analyses as are necessary to verify their validity and then to create the working drawings and specifications which will enable contractors to price and subsequently to construct the works. Procedures for controlling these operations include:

CALCULATIONS

The purpose of a calculations procedure is to ensure that calculations are produced in a consistent and methodical fashion, to minimize the risk of errors and to maximize intelligibility. It should identify the process by which the engineering judgment of senior design staff is brought to bear upon specific problems and the techniques adopted for their solution.

As an aid to designers, the calculations procedure should include a schedule of the approved computer programs in use within the office, identifying their purpose, capabilities and limitations. It may also include, or refer to, design charts, nomograms and conversion tables which have been developed within the office to cope with particular problems. It is essential that all calculations should carry an identification indicating the project to which they relate and brief details of their particular function. The procedure should provide examples of preferred presentation and should give instructions concerning numbering systems and arrangements for filing. Finally, the procedure should address the system to be used for checking calculations. This should include a mathematical check by another designer ('yellow line' check) and a review by senior staff for compliance with the design philosophy and final approval. Figure 6.2 illustrates a typical calculations check procedure and Figure 6.3 gives an example of a calculations check list.

DRAWINGS

This procedure should describe the routine operations of the drawing office. Its aim should be to provide a technically competent newcomer to the office with all the information he or she requires to become functionally effective with a minimum of additional instruction. It should establish standards of draftsmanship and give advice on matters such as layout, format and scales. Sections of the procedure should deal with drawing numbers, revisions, filing and issue.

The procedure should include a description of the system to be used for checking drawings before issue from the office. Figure 6.4 gives a flow chart

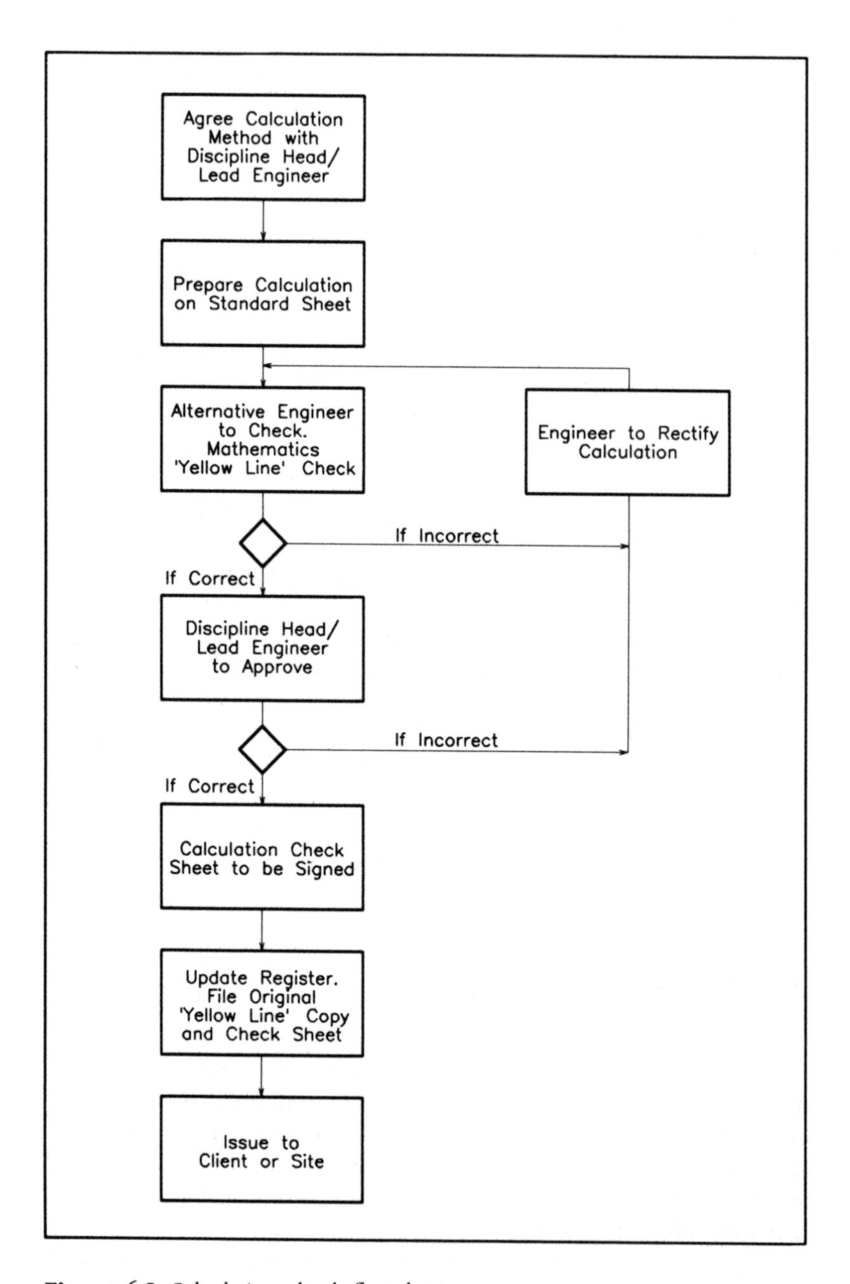

Figure 6.2 Calculation check flowchart.

A	CALCULATION CHECK SHEET		
CLIENT :			
CONTRACT :			
CONTRACT No. :			
CALCULATION No./REVISION :			DATE:
CHECKLIST	ORIGINATOR	CHECKER	DISCIPLINE HEAD OR LEAD ENGINEER
DESIGN PHILOSOPHY			
ASSUMPTIONS			
PRESENTATION			
NUMERICAL ACCURACY			
TITLE SHEET			
REFERENCES			
CONTENTS FORMAT			
INTRODUCTION			
LEGIBILITY			
DIAGRAMS			
UNITS			
TABLES			
SUMMARIES			
SHEET HEADINGS			
SHEET NUMBERING			
COMPUTER OUTPUTS			
APPROVAL	NAME	TITLE	SIGNATURE
ORIGINATOR			
CHECKER			
DISCIPLINE HEAD OR LEAD ENGINEER			

Figure 6.3 Calculation check list.

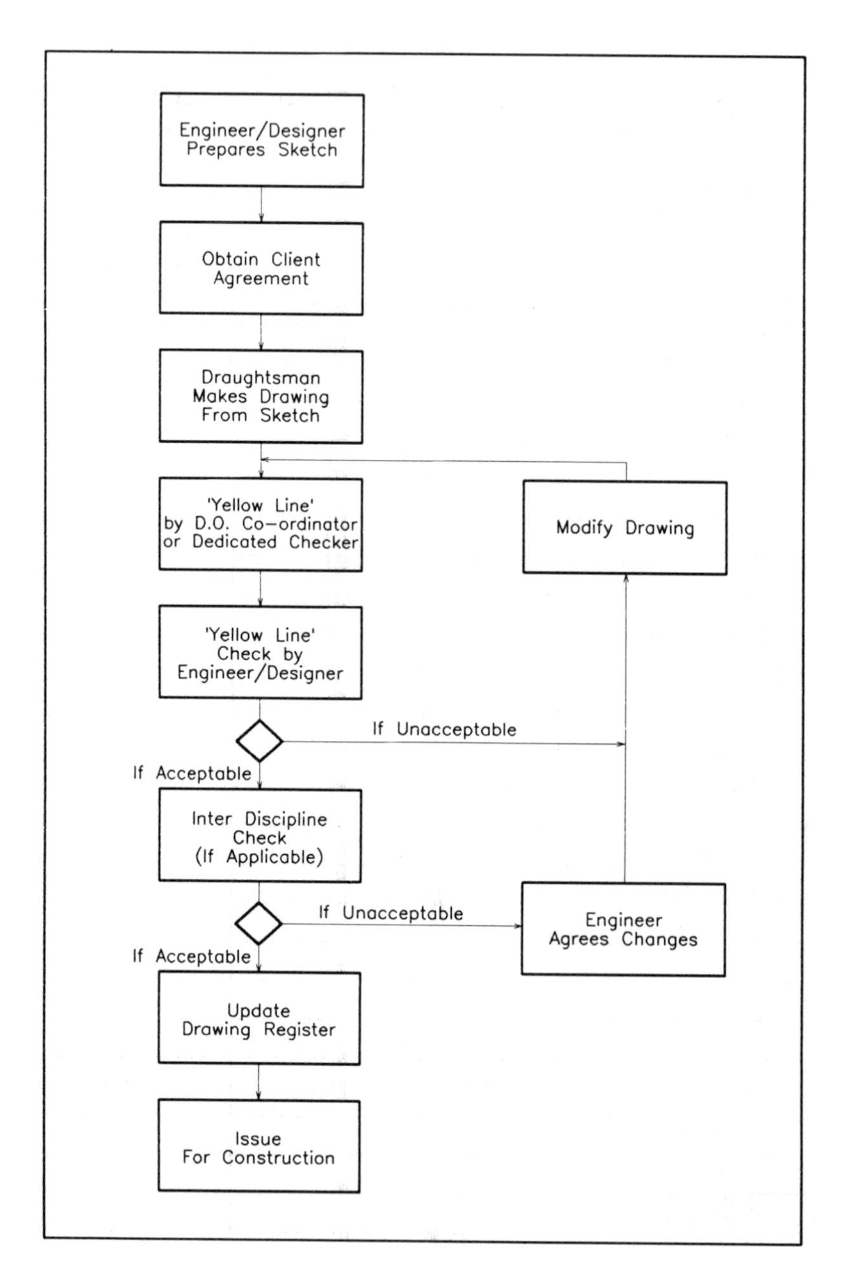

Figure 6.4 Drawing check flowchart.

A	DRAWING CHECK SHEET		
CLIENT : CONTRACT : CONTRACT No. : DRAWING No./REVISION :	DATE:		
CHECK LIST		**DRAUGHTSMAN**	**CHECKER**
DRAWING NUMBER			
TITLE BLOCK			
REVISION			
SIGNATURE			
LEGIBILITY			
SPELLING			
LAYOUT			
DIMENSIONAL ACCURACY			
NOTES			
REFERENCE DRAWINGS			
INTERPRETATION OF SKETCHES			
NEATNESS			
DRAUGHTING TO BS 308			
DRAUGHTING TO CLIENT PROCEDURES			
APPROVAL	**NAME**	**TITLE**	**SIGNATURE**
DRAUGHTSMAN			
D.O. CO-ORDINATOR OR CHECKER			
ENGINEER			
DISCIPLINE HEAD OR LEAD ENGINEER			

Figure 6.5 Drawing check sheet.

illustrating a checking sequence and Figure 6.5 gives an example of a drawings check list. Note that these examples apply only to drawings produced by a single discipline. If a number of disciplines are involved in a project, a further inter-disciplinary check is required as described later in this chapter under the heading 'Control of design interfaces'.

SPECIFICATIONS

Many design organizations maintain libraries of standard specifications covering the more common construction tasks and these are often applied unchanged to most projects. However, design engineers should examine each case on its merits and should test the validity of all standard specifications before they are issued.

Particular care should be taken when specifying construction tolerances as these are the cause of many quality problems. In the real world of construction, exact dimensions cannot and do not exist. Any dimension entered on a drawing will be subject to unavoidable imperfections in materials, measuring instruments, equipment and workmanship which, either singly or in combination, will result in a larger or smaller value in the finished work.

The specification of unreasonably tight tolerances can result in greatly increased costs without any compensating benefits. On the other hand, if the tolerances specified are very wide, or if none are specified, serious problems can arise on site from lack of fit. Given the extent to which construction work these days consists of the assembly of pre-manufactured components (particularly in building), this is a matter deserving of careful attention. The guiding rule should be always to specify the minimum standards that are acceptable. The practice of specifying high standards in the hope that they can be achieved, followed by an expedient relaxation when they are shown to be impossible or unreasonable is a mistake since it implies to those doing the work the notion that all other specification requirements are equally flexible and open to negotiation. What is more, the contractor of integrity who prices the work as specified will be undercut by his less scrupulous competitors who tender on the basis that lower standards will prevail, with or without the knowledge and agreement of the Engineer or Architect.

CHECK AND APPROVAL SIGNATURES

Evidence of control of design is normally provided by signatures or initials placed in the appropriate boxes of title blocks or check lists. By themselves, however, these marks are meaningless and do not provide sufficient verification of control for the benefit of auditors or of others who may be interested. There is therefore a need to validate signatures so that there can be no doubt that checks and approvals have been carried out by properly qualified and authorized persons.

Engineers delegated to assume professional responsibility for projects or sections of projects should select staff eligible to sign as checkers and approvers of documents on their behalf. They should maintain registers of authorized signatories and their signatures, identifying the categories of work within their competence.

Control of design interfaces

Most design projects require inputs from more than one technical discipline. There is a need, therefore, for a procedure to ensure that the work of each is compatible with all the others. This procedure should identify a communication system which will circulate all design teams with information on design proposals as they mature, so that conflicts or gaps can be detected and resolved in an orderly fashion.

Inter-discipline check (IDC) procedures require the preparation of distribution matrices for each discipline indicating the circulation requirements for the various categories of documents. A sample IDC circulation matrix is given in Figure 6.6. To control the circulation of documents, it is normal practice to establish a document control centre whose purpose is to expedite and keep track of movements. After carrying out their internal checks, originating disciplines supply the document control centre with an original of the document to be circulated. According to the circumstances and degree of urgency, the document control centre may then either circulate a single copy on a series routing to every interested discipline, or print a number of copies and make a parallel issue. Comments received from the IDC circulation should be reviewed by the originating discipline. Further discussion between disciplines may be necessary to resolve disagreements. In the event of changes in principle, revised drawings should be re-submitted to the document control centre for a second IDC.

Change control

It is a rare event indeed for a design document to remain unchanged throughout its lifetime. More often it will be issued in a succession of modified editions. There is a variety of reasons why this should be so: perhaps circumstances have changed, equipment may have become obsolete, someone may have thought of a better way, or it may simply be that the purchaser has changed his mind. Whatever the reason, it is vital that

A	INTER DISCIPLINARY CHECK (I.D.C.) DISTRIBUTION MATRIX	SHEET of
		Date

CIRCULATED DISCIPLINES / ORIGINATING DISCIPLINE	PROCESS	PIPING	MECHANICAL	INSTRUMENT	ELECTRICAL	LOSS PREVENTION	STRUCTURAL	DRAWING OFFICE	COMPUTING	PROJECTS	ENG. MANAGER	QUALITY ASSURANCE			

REV.	DATE	BY	APPROVED	REMARKS

Figure 6.6 Interdisciplinary check distribution matrix.

the correct editions of documents, and only the correct editions, should be used for the performance of any given task.

Change control commences with the control of incoming documents. These will include the client's brief, and the standards, codes of practice and statutory regulations with which design work has to comply. The system should ensure that every copy of these documents can be located so that the new can be substituted for the old, apart from a master copy of each edition which should be kept for record purposes. Beware, however, of contracts which specify particular editions of standards. The correct edition in such cases may not necessarily be the latest edition.

Drawings and specifications originating from within the office require similar control procedures. Evidence of control is provided by the data supplied in title blocks and in Document Registers which should record issues and recipients. Note that amended documents should undergo the same checking processes as the originals, and that they should be checked and approved by properly authorized persons.

Design reviews

In addition to the routine control activities described above, it is good practice to institute a programme of periodic reviews by persons outside the project team to ensure that design work is proceeding on the right lines and that the objectives defined in the brief are being achieved. It is frequently the case that those who are intimately concerned with the minutiae of design will overlook important matters which happen not to appear within their fields of vision but which are immediately apparent to an experienced outside observer.

Design reviews may be undertaken by higher tiers of management, by peers of the project team or by specialized review panels. There are significant advantages if the client can be persuaded to take part. Typical questions addressed within design reviews include:

1. Are the design techniques which are being used appropriate for the particular type of work?
2. Are members of the design team suitably trained and qualified for the tasks they are required to perform?
3. Have all factors relative to the design been taken into account?
4. Do the designs satisfy health and safety requirements?
5. Are the designs 'buildable'?

To speed the conduct of design reviews, project teams may be required to make formal presentations of their work. This is a most valuable discipline as it obliges the team to marshal its thoughts in advance and to anticipate the questions likely to be asked. This exercise can be beneficial in itself and may bring to light important issues even before the design review is held. The results of design reviews should be documented.

Computer software

The quality assurance of computer software is a topic about which much has been written. It is a subject requiring considerable specialist expertise, most of which is outside the scope of this book. The paragraphs which follow deal only with routine procedures for the control of software which are appropriate to a design office serving the construction industry. Computer applications in design offices may be categorized as follows:

1. *Administration* Examples include the processing of time sheets, network planning, accounting systems and the payment of wages and salaries.
2. *Equipment* Equipment incorporating software includes computer-aided drafting machines, digitizers and word processors.
3. *Computation* This category comprises the software used by designers for analysis and calculations.

Software for categories 1 and 2 will not be discussed in detail because of its specialist nature. The most important category for design engineers is Category 3, since this is likely to have the greatest influence on design quality. Let us consider the principal hazards which a computation software control procedure should be designed to prevent:

Use of unapproved programs
Defects (bugs) in software
Erroneous input
Misinterpretation of output.

In well-established design offices, the purchase, custody and administration of computer software and hardware are likely to be the responsibility of a nominated manager. Let us call him the Computer Systems Manager. He, the discipline heads, the project engineers and the project staff will combine to implement the software control procedure which should be subject to regular audit by the Quality Manager. The following is a typical allocation of duties:

Discipline heads

1. Advise Computer Systems Manager on the specification and purchase of new programs.
2. Ensure the technical integrity of programs, including the soundness of their theoretical basis.

3. Ensure the adequacy and completeness of documentation.
4. Make recommendations for modifications.
5. Establish training programmes for users and verify their competence.

Computer systems manager

1. Maintain a register of approved programs.
2. In collaboration with discipline heads, specify and purchase new programs.
3. Test and approve the originals and all modifications of programs and accompanying documentation.
4. Write new programs when requested by discipline heads.
5. Maintain archive records of superseded program revisions.
6. Control, maintain, up-date and issue User Manuals.
7. Investigate bug notifications and take appropriate actions.
8. Provide a security system to prevent the irresponsible or illicit use of computer facilities.

Project engineers

1. In consultation with discipline heads, establish the schedule of programs approved for project use.
2. Apply to Computer Systems Manager for the provision of computer facilities and a Project User Identification.
3. Nominate project staff to have access to project computer facilities and advise them of passwords and logging-on procedures.
4. Authorize major computer runs.
5. Maintain computing log for project.
6. Protect User Identification, passwords, programs and data against unauthorized disclosure.
7. Review or re-validate User Identifications when necessary and change access passwords in accordance with security procedures.
8. Report program defects (bugs) to Computer Systems Manager.
9. Ensure that appropriate checking procedures are followed.

Project staff

1. Follow User Instructions in operating programs.
2. Carry out manual checks to verify computer calculations.
3. Report program defects (bugs) to the Project Engineer.

Note that most of the activities scheduled above are directed either to preventing the use of unapproved programs or to the elimination of program defects. Important though these measures are, it needs to be borne in mind that design errors are more likely to be caused by incorrect input data or the misinterpretation of output than by defective or inappropriate software. The only means of preventing such errors is the employment of designers who have a sound engineering training which equips them to understand the manipulations which the computer is carrying out on their behalf and enables them to check by manual calculations, or by instinct, that the answers to their calculations are correct.

Feedback

Quality in design requires that designers should receive adequate feedback on the realization and practical performance of their designs. In construction work, this feedback may be of three kinds:

1. The designer needs verification that the criteria upon which the design is based (for example, on ground conditions) are valid.
2. The designer needs to satisfy himself that the works are in fact constructed in accordance with his instructions.
3. The designer needs to be aware of the performance of his designs, both during the construction period and in their finished state. Without such knowledge, errors will be repeated and the steady improvement and refinement of design technique which are essential to success will not take place.

To obtain this feedback, the designer requires unimpeded access to the construction site. One of the principal advantages of traditional contractual arrangements is that the rights and obligations of the Architect and the Engineer to participate in the construction process are established in the contracts. Any restriction of these rights should be resisted.

Records

The general points contained in Chapter 5 on the subject of records apply as much to design as they do to construction. Design offices maintain records for the following purposes:

1. To satisfy clients' requirements for construction details.
2. To provide evidence in substantiation of requests for payment.
3. To preserve knowledge of designs and techniques for use on future projects.
4. For possible presentation in rebuttal of claims for compensation in respect of negligent work.

The purposes for which records are maintained will influence the times of retention and each item needs to be considered on its merits. Microfilming of records achieves savings in space, but the rules governing admissibility as evidence in legal proceedings need to be observed. (See p. 85.)

7

PROCUREMENT

Introduction

It is difficult to imagine any commercial organization which is so self-sufficient that it does not need to purchase materials, components and services from others. Such is the interdependence of modern society that we are all buyers and sellers, and every company relies on the skills of its procurement department to ensure an inward flow of defect-free goods and services, on time, and at the right price.

In the construction industry, the traditional procurement sequence commences with a primary purchaser, usually referred to in contracts as the Employer. The Employer purchases design and site supervision from an Architect or Engineer and places a contract for construction work with a contractor. The contractor purchases materials and components from suppliers for incorporation into the works and may also place sub-contracts for parts of the construction work with other, usually smaller, contractors. Some of the sub-contractors may also purchase construction materials, and they may further sub-let parts of their sub-contracts, and so on. There is thus a wide range of interlocking transactions, embracing many different parties, all of which have to be controlled if the Employer's needs for a satisfactory project are to be fulfilled. The same variety of transactions, although perhaps in a different sequence, will also take place if the Employer opts for non-traditional methods of procurement such as management contracting, or a design-and-construct contract.

The one common feature in all these transactions is that a commodity is purchased. The commodities may be completed projects, such as a length of motorway, or manufactured articles like pumps or bricks, or the performance of services where there is no visible end-product except perhaps a sheaf of drawings or a test certificate. Writers of quality system standards appear to have had difficulty in finding a suitable generic term capable of encompassing the whole range of things bought. Standards derived from the early association of quality assurance with defence procurement, such as DEF STAN 05-21 and early versions of BS 5750, used the term 'materiel'. This is a word of French origin, used to describe the baggage and munitions of an army. BS 5882 takes a simpler line. It uses the word 'item' which it defines thus:

Item: An all-inclusive term covering structures, systems, components, parts or materials.

These definitions may be adequate for material purchases, but they do not cover contracts for services which are by their nature intangible. BS 4778 does not define 'services', but the following is from an American specification, *'Quality Assurance Program Requirements for Nuclear Facilities, ANSI/ASME NQA-1'*:

Services: The performance of activities such as design, fabrication, inspection, non-destructive examination, repair or installation.

The 1987 version of BS 5750 solves the problem quite satisfactorily by dispensing with definitions. It simply uses the word 'products' and explains in a preamble that '..."product" is also used to denote "process" or "service", as appropriate...'. For convenience, the same convention will be used in this chapter.

The principal features of procurement quality systems are the same for all buyers of products, whether they are primary purchasers, such as the Employer in a construction contract, secondary purchasers (main contractors) or tertiary purchasers (sub-contractors or material suppliers). They are:

Selection of potential suppliers
Definition of requirements
Verification of compliance.

Let us consider each of these functions in turn as they may be implemented at different stages in the construction sequence.

Selection of potential suppliers

It makes good sense to ensure that suppliers of products should have both the means and the will to comply with the requirements of an order. In the construction industry it has long been the custom for contractors to be obliged to provide information about their financial standing, their resources of people and equipment, and their record of past work in order to pre-qualify for inclusion on tender lists. An effective quality management system will similarly require a demonstration of a supplier's capability to control the works and to assure conformance with specified requirements. In some quality system standards, potential suppliers are referred to as 'vendors'. A 'vendor' becomes a 'supplier' when a purchase order has been issued or a contract has been signed.

BS 5750:Part 0.2 lists the following methods of establishing a vendor's capability:

(a) on-site assessment and evaluation of supplier's capability and/ or quality system
(b) evaluation of product samples
(c) past history with similar supplies
(d) test results of similar supplies
(e) published experience of other users.

APPROVED LISTS

Many purchasing departments use the above methods to assemble an 'approved list' of potential and existing suppliers and sub-contractors. These lists may be based on a simple card-index system or, in larger organizations, they may be maintained on a computer database to provide rapid updating of information and easy access to a number of users. Approved lists may contain:

1. Supplier's name and address.
2. Contact name and telephone number.
3. Commodities or services for which approval has been granted.
4. Details of current orders.
5. Details of recently completed orders.
6. Feedback reports from users.
7. Details of recent company pre-contract assessments or other audits.
8. Registration by accredited third-party certification bodies.

PRE-CONTRACT ASSESSMENT

A pre-contract assessment is a formal procedure to determine a potential supplier's eligibility for inclusion on an approved list or to tender for a specific order. An assessment is a form of quality audit, the techniques of which are dealt with in detail in Chapter 9.

Before embarking on a pre-contract assessment, a purchaser normally seeks preliminary information by means of a questionnaire such as that illustrated in Figure 7.1. If this indicates that a formal assessment is justified, the next step will be a more detailed examination of documentation submitted by the supplier. The documentation required will be specified by the purchaser together with identification of the quality system standard (if

A QUALITY SYSTEM QUESTIONNAIRE

NAME OF ORGANISATION

ADDRESS:

TEL No.

SERVICES OR GOODS OFFERED BY YOUR ORGANISATION:

Does your organisation: DESIGN ☐ MANUFACTURE ☐
INSTALL ON SITE ☐ CONSTRUCT ON SITE ☐ (Tick as applicable)

1. Does your organisation have a documented Quality System? If YES, state to which Standard it relates (if any) — YES/NO

2. Does your organisation have a Quality Manual? If YES, please attach a photocopy of the index or contents page to this Questionnaire when it is returned. — YES/NO

3. Does your organisation have other Manuals or Procedures which relate to the control exercised over its quality management activities? If YES, state titles: — YES/NO

4. Does your organisation have a person appointed as "Quality Manager" (or similar title)? If YES, please state his name, functional title and to whom in the organisation he reports: — YES/NO

5. Does your organisation have a current formal approval or registration in respect of it s quality system to a national or industry sector scheme? If YES, give details with Approval/Registration No. and effective date: — YES/NO

6. Has your organisation received any quality system audits by any major companies within the last 12 months? If YES, give brief details and dates: — YES/NO

Signature Job Title Date

Figure 7.1 Quality system questionnaire.

any) against which the audit will be made. Typically, the documentation requirement will be satisfied by the supplier's quality manual, possibly supplemented by additional material specific to the product or contract under contention. Any significant omissions or deviations from the standard are notified to the supplier so that appropriate amendments or improvements can be made. The appraisal of documentation is then followed by a visit to the supplier's factory, head office or site to confirm that the description of the quality system given in the manual is factual and can be supported by evidence.

Assessments carried out by purchasers are known as second-party assessments. These can be very expensive, both for purchasers and suppliers. Even a very cursory assessment will require the time of two or three persons for one or more days and major audits may involve teams in excess of ten people for periods measured in weeks. Suppliers can find themselves continually being assessed by potential purchasers, with only a small minority of assessments leading to actual orders. Likewise, it can be a major task for a purchaser to assess all suppliers with whom he may wish to do business. This wastage of resources can be reduced by the practice of third-party assessment.

Third-party assessments are undertaken by independent bodies established for the purpose. They carry out audits and surveillance following procedures similar to those of second parties and award a mark or registration to suppliers who succeed in meeting their standards. Provided the third-party organization has recognition and status in the eyes of purchasers, the mark or registration it awards will eliminate the need for second-party assessment and surveillance as long as it remains valid. Organizations undertaking third-party assessment and surveillance are known in the United Kingdom as 'certification bodies'. Some, such as the British Standards Institution, have for many years carried out third-party product certification (the 'kitemark' scheme is an example of this) and, for these, an extension of their operations into the certification of companies is a logical step. Other certification bodies have been formed on the initiative of groups of companies within specific sectors of industry who have perceived a need for independent certification. These moves by industry have received substantial encouragement from government through the provision of financial support and the establishment of the National Accreditation Council for Certification Bodies (NACCB). This organization is responsible for publishing guidelines for the proper constitution of certification bodies covering such areas as technical competence, financial probity, consumer representation and so on, and thereafter for examining applications and accrediting those which comply. The criteria against which it makes its judgments are based on harmonized international guidelines to encourage reciprocal recognition of certification arrangements.

There are four categories of certification for which aspiring bodies may seek NACCB accreditation:

1. Certification of quality management systems
2. Product conformity certification
3. Product approval
4. Certification of personnel engaged in quality verification.

Of the above categories, the first two have the most significance for the construction industry. Bodies with category 1 certification (known as 'assessment bodies') assess and register the quality systems of suppliers with respect to published criteria, such as BS 5750. Their functions do not include the certification of products, processes or services and they do not carry out any checks to verify that these comply with specifications. Organizations to whom registration is granted receive a certificate and become entitled to display the certification body's symbol or logo, subject to certain restrictions. They may not, for example, use the symbol or logo on a product, or in a way that may be interpreted as denoting product conformity.

Certification bodies accredited for product conformity certification go much further. In addition to assessing quality systems, they take representative samples for both initial and subsequent audit testing and they arrange for testing to be carried out in suitably accredited laboratories. Not only do they issue registration certificates, they also grant permission for their symbols or logos to be displayed on the recipient's products. The United Kingdom Certification Authority for Reinforcing Steel (CARES) is an example of a certification body accredited to assess suppliers for product conformity.

Pre-contract assessments will frequently reveal that some potential suppliers do not have quality systems which satisfy quality system standards, or indeed may not have any identifiable quality systems at all. This need not necessarily disqualify them from tendering for work. There would be very few bricklaying sub-contractors on construction sites if all had to operate documented quality systems. A supplier who has performed satisfactorily on similar work in the past, or who is able to demonstrate that his product consistently meets specification requirements can justify inclusion on an approved list on these grounds alone.

Even after taking all the actions outlined above, a purchaser may still be faced with a predicament in which an unapproved vendor submits a tender substantially below those received from approved vendors. In a competitive situation, such offers are difficult to ignore. The correct response is to add to the unapproved vendor's offer a sum of money which will provide sufficient additional supervision by the purchaser himself to give a degree of confidence equivalent to that demonstrated by the other vendors. If his offer is still the lowest, he is entitled to the work.

A

SUPPLIER / SUB-CONTRACTOR
PERFORMANCE SUMMARY

SUPPLIER/SUB-CONTRACTOR:

ADDRESS:
TELEPHONE: TELEX:

PROJECT/CONTRACT:

PURCHASE ORDER No. & DATE:

MATERIAL/SERVICE SUPPLIED:

RATING
A = Acceptable
B = Some improvement required
C = Substantial improvement required
D = Unacceptable

WAS:		A	B	C	D
– a quality plan submitted when required	1				
– work carried out in accordance with an approved quality plan	2				
– quality of all work acceptable	3				
– delivery and completion date met	4				
– operational documentation satisfactory	5				
DID SUPPLIER/SUB-CONTRACTOR:					
– co-operate with supervisors	6				
– react effectively to quality problems	7				
– offer a good standard of technical co-operation	8				
– react effectively in emergencies	9				
– maintain technical representation in the field	10				
– advise us promptly of potential trouble	11				
– efficiently handle introduction of any new or unfamiliar materials or processes	12				
– adequately control his own suppliers/sub-contractors	13				
– adequately protect and preserve finished work and products	14				

COMMENTS (continue on separate sheet if necessary)

Signature Job Title Date

Figure 7.2 Supplier performance summary.

FEEDBACK

The final test of a supplier's competence is his ability to deliver the goods. This requires a reporting routine whereby the users of products or services are able to record the degree of satisfaction they have received. A typical standard form for providing this information is illustrated on Figure 7.2.

Definition of requirements

If a purchaser cannot, or will not, define what he wants, he has little cause for complaint if the products he receives fail to satisfy his needs.

Apart from private housing, it is only rarely that buyers of construction products can see their purchases in their finished state before committing themselves to spend money. One cannot look at dams, or bridges or office blocks in a showroom and decide which model to buy. It is therefore incumbent on the purchaser of construction products to identify his needs and communicate them to his designers and suppliers in an unambiguous fashion. The normal means of communicating the Employer's requirements to a contractor is by way of technical specifications and drawings. However, before these can be prepared, the designer has to be aware of the Employer's needs so that he can embody them in his designs in a manner which merits the Employer's approval. Chapter 6 discussed this subject in some depth.

As the Employer's requirements move down through the stages of procurement, they tend to acquire an accretion of detail as each participant adds his particular requirements. For example, a primary purchaser might instruct his designer that he requires a pump to extract fluids from a sump. The designer will determine the type and characteristics of the pump and pass these to the contractor. The contractor will select the supplier, and confirm to him the technical specification, together with requirements for stage inspections, packing, spares, delivery dates etc. The pump manufacturer will specify the detailed requirements of each component and pass these to his suppliers and so on. There is thus a proliferation of detailed information much of which does not concern the primary purchaser, but some of which may. Employers deal with this problem in different ways. At one extreme there are those who require copies of every purchase order, sub-order and sub-sub-order so that they may maintain tight control at each stage. Others take the view that, having selected their suppliers and given them a contractual responsibility, the suppliers should be left to sort out the details for themselves.

THE REQUISITION

Definition of the requirement normally commences with the preparation of a document known as an indent or requisition. This provides a description of the materials, equipment or services required and forms an instruction to the buying staff to obtain quotations or to commit the expenditure of money. The preparation of requisitions should be a technical function, and in the case of substantial projects should be delegated to suitably qualified personnel within the project team. Organizations undertaking a number of small projects may alternatively maintain a head office team for the task. For the purchase of conventional building materials, the requisition may comprise a simple single-page document pre-printed with spaces for the project name and number, a materials schedule, the specification requirements, the required delivery date and the signatures of the originator and the site agent. Such indents may be 'for enquiry', in which case the buyer is expected to obtain quotations and revert to the originator to seek authority to purchase, or they may instruct the buyer to go ahead and purchase on his own initiative.

Complex equipment calls for a more comprehensive procedure. There may be a requirement for a technical review of bids as well as a commercial assessment, and there may also be a need to obtain the Employer's approval of the proposed purchase. In such cases, the requisition will grow to a multi-page dossier and pro-formae to incorporate the following will be required:

1. A general description of the materials or equipment to be purchased or the service required.
2. Schedule of recommended suppliers (if available).
3. Delivery date(s).
4. Budget value.
5. Shipping instructions.
6. Signature of approving authority.

The above will be attached to a Materials List and a Specification Package. The Materials List will schedule the quantities and items to be purchased and give reference to the appropriate specification requirements. The Specification Package may include the following:

1. A schedule of standard specifications to be complied with.
2. Copies of project specifications.
3. Drawings.
4. Data sheets.
5. Audit, inspection and certification requirements.
6. Requirements for suppliers' documentation (specifications, drawings, operating manuals, etc).

7. Packing specification.
8. Tagging requirements.
9. Spare-part requirements.

THE PURCHASE ORDER

The evolution of a requisition into a purchase order is illustrated by the flow chart on Figure 7.3. In this example, the Enquiry Requisition is passed to the Purchasing Department where it is combined with appropriate commercial documentation to form an Enquiry Package capable of being priced by potential suppliers. Tenders received in response to the Enquiry Package undergo commercial analysis and, if necessary, a technical review and are returned to the originator together with summaries and recommendations.

The originator's task is then to decide upon the most acceptable offer and to prepare a 'for purchase' version of the requisition incorporating any approved amendments arising from the chosen supplier's offer. The flow chart on Figure 7.3 shows that before transmittal to the Purchasing Department, the requisition is subjected to a quality assurance review. The quality assurance function will wish to satisfy itself that the requirements for documentation, surveillance and audit are adequately catered for and that the procedures for pre-qualifying and selecting suppliers have been adhered to. Quality assurance reviews are sometimes also made of enquiry requisitions, particularly if there is a likelihood that the quality assurance requirements might have a significant effect on the tenderer's quotations. This is unlikely in the case of most standard building materials, but when more complex purchases are concerned the costs of documentation, surveillance, records, etc. may be considerable. There may also be a requirement that enquiries should be solicited only from suppliers on an approved list or who have been assessed by an accredited certification body.

A rule that all requisitions must be reviewed by the quality assurance function both at the enquiry and purchasing stages can be a blunt instrument. Such arrangements can inadvertently contribute to the misapprehension that quality management is a paper-chasing ritual. It is better to have a procedure which enables requisition originators to select those which require a quality assurance review by using pre-determined criteria. These may be based on the category of work, the value of the order or the commodity being purchased. Such criteria should form part of the Project Quality Plan.

The approved purchase requisition finally reaches the Purchasing Department where it becomes the basis of the formal Purchase Order to the supplier.

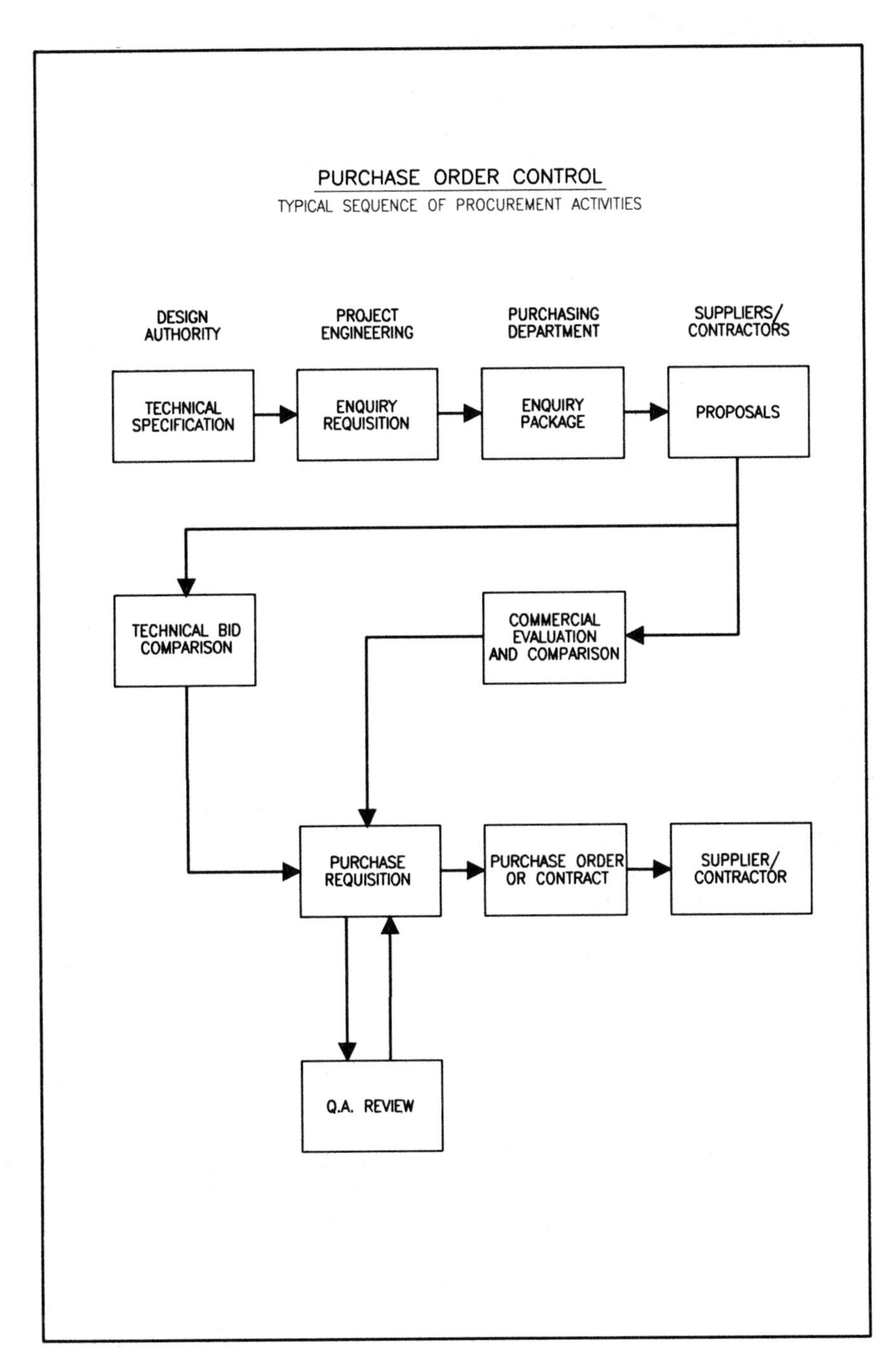

Figure 7.3 Purchase order flowchart.

Verification of compliance

One of the functions of a Purchase Order is to define the procedure to be followed to give confidence to the purchaser that the specification requirements relating to the purchased products are complied with. There is a number of options, which includes:

1. The purchaser may specify that the supplier implements a particular formal quality assurance system.
2. The purchaser may rely on the supplier's own quality system.
3. The purchaser may rely on the supplier's reputation.
4. The purchaser may require the supplier to provide documentary evidence of compliance.
5. The purchaser may institute his own inspecting and testing regime, either in place of, or in addition to, that of the supplier.

Let us consider examples of the above five options as they may be exercised in the construction context.

OPTION 1

An electricity generating board wishes to construct and operate a nuclear power station. The national licensing authority requires that it should comply with a specification for a quality assurance programme for nuclear installations both in respect of its own operations and its suppliers and contractors. The board therefore makes it a condition of contract that all concerned will comply with the quality assurance specification and defines in detail the steps which the contractors will have to take in order to do so. The board gives itself the right to approve or disapprove quality documentation, to audit contractors' systems and to impose corrective actions.

OPTION 2

A consulting engineer arranges for a testing laboratory to carry out a series of tests on soil samples taken as part of a site investigation. The laboratory's quality system has been assessed by the National Measurement Accreditation Service (NAMAS) and has been granted their registration in respect of the tests to be carried out. The consulting engineer recognizes NAMAS accreditation as an adequate demonstration of the laboratory's capability, and concludes that the test results supplied may be relied upon.

OPTION 3

A lay client engages an architect to design a building on his behalf. The selection is made on the basis of previous work undertaken by the architect for the client, and on the architect's high reputation in designing buildings of the type required. The architect signs an undertaking that he will 'exercise reasonable skill and care in conformity with the normal standards of the architect's profession' in designing the building and supervising its construction. The client takes it on trust that the architect will operate a quality system appropriate to the work and that the resultant building will be to his satisfaction. The architect has not offered fitness for purpose, and there are no contractual sanctions available to the client in the event of default other than termination of the engagement after a reasonable period of notice. The entire deal rests upon the architect's professional skill and integrity. The client relies upon this as the sole source of confidence that his needs will be met.

OPTION 4

A contractor issues a Purchase Order to a supplier of cement. The order specifies the type and quantity of cement and the standards with which it is required to comply. It also requires that the supplier should forward a test certificate confirming that at the time of delivery the cement complied with the requirements of the standard and recording the results of physical tests carried out on samples taken from the consignment.

OPTION 5

An Employer engages a contractor to construct a bridge. The contract requires that the work should be in accordance with drawings and specifications provided by an appointed Engineer acting on behalf of the Employer. It also requires that the contractor should carry out tests on materials and workmanship as specified and paid for. An Engineer's Representative is appointed to watch and supervise the works and to ensure that the specified testing regime is adhered to. He is empowered to reject materials which are not of the required standard and to order the removal and re-execution of work which is not in accordance with the contract.

The activities on the site of the Engineer and his representative do not detract from the contractor's responsibility to build the bridge in accordance with the drawings and specifications, and to carry out the necessary tests. Their function is the provision of confidence to the purchaser that the contractor is, in fact, carrying out his contractual duties and that the works are, in truth, as specified.

Criticality rating

The range of options exemplified above demonstrates that the degree of confidence required by purchasers varies according to circumstances. A purchaser who in one case will demand total and unshakeable proof of compliance with specification may in others be quite prepared to accept a supplier's word that he has performed as required. A quality system should be able to accommodate these differences. It implies no compromise on the obligation to comply with specification to conclude that in any project there will be elements of work which will differ from one another in their importance with regard to safety, reliability, complexity and so on. A practical balance has to be struck between the costs of the various measures which may be taken to assure quality and their benefits.

'Criticality Rating' is a formalized technique for measuring the importance to be assigned to an element of work or piece of equipment as measured by the consequences of its failure. There are many criteria by which criticality may be judged and some organizations operate procedures which allocate points for various aspects such as safety, stand-by availability, the financial consequences of failure, access for replacement and so on. The points allocated to each element or area of work are then used to determine the intensity of the control measures to be applied.

As an example of the application of a criticality rating system, let us consider the case outlined in Option 1 on p. 118. In the United Kingdom, the national standard for nuclear quality assurance is BS 5882 *A total quality assurance programme for nuclear installations*. This sets out a comprehensive series of rules for assuring quality to bring about a safe and successfully operating installation. However, as was discussed in pp. 43 and 44, not all elements of a nuclear facility are critical in terms of safety. The standard therefore allows alternative methods or levels of control and verification and offers guidelines as to how these should be graded.

In the nuclear industry, the word 'criticality' carries ominous overtones arising from its use to describe a particularly dangerous stage in a nuclear reaction. In the discussion of this example the words 'safety grading' will be used instead of 'criticality rating'. They describe the same process without any hidden meanings, and serve to demonstrate that the paramount (although not only) purpose of nuclear quality assurance is the achievement of safety. In the public mind the possibility of a major nuclear accident is regarded as a uniquely horrifying event. The most stringent category in a safety grading system must therefore encompass components or systems, the failure or lack of availability of which might contribute to such an accident. Let us call this category 'Q'. These items will require the full application of BS 5882.

Table 7.1 Typical grading of control and verification

Control and verification activities	Grading			
	Q	SO	E	N
1. Submit QA programme complying with BS 5882	Ap	No	No	No
2. Submit documented QA System, e.g. to BS 5750	–	Ap	Inf	No
3. Submit quality system procedures	Ap	Ap	Inf	No
4. Submit quality plans	Ap	Ap	Inf	No
5. Submit sub-contractor quality plans	Ap	Inf	No	No
6. Quality system monitored by:				
Audit	Yes	Yes	No	No
Surveillance	Yes	Yes	Yes	No
Source inspection	Yes	Yes	No	No
Receipt inspection	Yes	Yes	Yes	Yes
7. Design details required	Ap	Ap	Inf	No
8. Fabrication details required	Ap	Ap	Inf	No
9. Construction details required	Ap	Ap	Inf	No
10. Notification/Hold points pre-specified	Yes	Yes	Yes	No
11. Submit deviation/concession notices	Ap	Ap	Ap	No
12. Quality release for shipment to site	Yes	Yes	Yes	No
13. Lifetime record data package required	Yes	Yes	No	No

Key: Ap: For approval; Inf: For information; Yes: Item required; No: Item not required.

Next there will be aspects of the work which while not having a direct effect on nuclear safety, may either have serious non-nuclear safety implications or may have a major influence on the operational functioning of the plant. Let us allocate these to category 'SO'.

A further category comprises items which, while not critical in terms of safety or operations, could nevertheless cause serious delays or cost overruns. This would apply where there is innovation, or the use of non-standard equipment or cases where complexity of design or manufacture is likely to cause trouble. Call this category 'E'.

Finally there are products which have negligible effect on the safety or operational efficiency of the plant and where normal commercial standards will be adequate. These items will be in category 'N'.

It is the responsibility of the owner of the proposed plant to allocate gradings and to inform potential suppliers of the verification requirements applicable to each of the products or services to be supplied. Table 7.1 is an example of a purchaser's schedule of requirements for quality control and verification in respect of the four safety gradings detailed above. This would be issued to tenderers together with a matrix or table identifying the allocated grading of each item to be priced (Table 7.2). As part of the prequalification process, tenderers would be called upon to demonstrate their ability to comply with the requirements of the most stringently graded item contained in the respective work packages, and would thereafter be expected to allow for the costs of meeting these requirements in their tender prices.

Table 7.2 Typical grading for a nuclear installation

Item	Building Group		
	1	2	3
1. Excavation	N	N	N
2. Backfill	Q	E	E
3. Drainage	SO	E	N
4. Piling	E	N	N
5. Concrete	Q	E	E
6. Reinforcement	Q	E	E
7. Pre-stressing	Q	SO	E
8. Structural steel	Q	SO	E
9. Welding	Q	SO	E
10. Embedments	Q	SO	N
11. Pipe supports	Q	SO	N
12. Fireproofing	SO	E	N
13. Brickwork	SO	N	N
14. Carpentry	N	N	N
15. Asphalt roofing	E	N	N
16. Partitions	N	N	N
17. Cladding	SO	E	N
18. Plastering	N	N	N
19. Glass and glazing	N	N	N
20. Painting	SO	N	N
21. Sanitation	N	N	N

Key to Building Groups: 1 Nuclear Island; 2 Turbine House, Cooling Water System, Switch Yard; 3 Other Buildings.

8

THE CONSTRUCTION SITE

The quality management of work on the construction site takes place in three phases:

1. Planning what is to be done.
2. Controlling the execution of the plan.
3. Providing verification that the work has been carried out according to plan.

Let us consider the requirements of a quality system in the context of these phases.

Planning

The steps to be taken to ensure that specified standards are met need to be planned in a systematic fashion and they have to be taken into account when overall work plans are being prepared. Design plans, constructions plans, cost plans, and so on, are part of everyday site management. So, too, should be quality plans.

CONTRACT REVIEW

Before starting work, the contractor should make sure that his client's requirements are clearly understood, and should clear up any ambiguities or contradictions. He should also make sure that his organization is equipped to do the work. Most contractors carry out a review of this kind before tendering. It makes sense to repeat it, with the co-operation of the client or his representative, before starting work. It is also important to keep appropriate records of agreements made.

QUALITY PLANS

Chapter 3 (p. 50) quoted the following international definition from BS 4778:Part 1.

Quality Plan A document setting out the specific quality practices, resources and activities relevant to a particular process, service contract or project.

BS 5750:Part 0.2 further advises that quality plans should define:

1. The quality objectives to be attained;
2. The specific allocation of responsibility and authority during the different phases of the project;
3. The specific procedures, methods and work instructions to be applied;
4. Suitable testing, inspection, examination and audit programmes at appropriate stages (e.g. design and development);
5. A method for changes and modifications in a quality plan as projects proceed;
6. Other measures necessary to meet objectives.

The above headings bear a close resemblance to those of a Quality Manual as discussed in Chapter 5. This is predictable since the purpose of a quality plan is to give effect to the company quality system in the context of a particular project. Note that in work governed by BS 5882 *A total quality assurance programme for nuclear installations* and in the petrochemical and offshore fields, such documents are commonly known as 'quality programmes'.

To be of value, the first issue of a quality plan must be made before the commencement of work on site. It is also essential that it should be a document which will be read, valued and used by those in control of work. These two statements may be considered self-evident, but it is an unfortunate fact that quality plans are all too often late, overweight and disregarded by those whose efforts they are meant to assist. Their preparation should be commenced at the tender stage as part of the normal routine of construction planning. It should not be an additional ritual imposed on the site team by outside agencies.

The quality manual of Alias Construction Ltd, which is reproduced as Appendix B, outlines the company's policy on the preparation of quality plans (para 5.3). More detailed instructions are given in Alias Construction's Company Standing Instruction CSI OPS-03 'Quality Plans'. This instruction is not reproduced in this book, but an example of a quality plan assembled in accordance with its instructions is included as Appendix D. The fictitious project to which the plan relates is the construction of a small-to-medium-sized commercial building and the example is presented as a typical first post-tender version of a project quality plan. The main purpose of the quality plan is to focus the minds of those in charge of work upon the prevention of defects. The example in Appendix D would require perhaps one day's work by the three persons concerned. Its preparation would involve the examination of drawings and specifications to identify problem areas and the allocation of resources and time for devising and documenting the methods to be used to overcome them. The plan also requires that decisions should be made on 'Who does what?' and on how the various checks and inspections are to be recorded. Most people would agree that these are all examples of sound construction management practice.

At the time of the first issue of a project quality plan, many of the problems brought to light will not have been resolved. Others will not yet have emerged, since it is rare for the design of a project to be completed before construction starts. It is therefore necessary that the plan should be updated as the project proceeds, with regular reviews to check that the planned activities are in fact taking place and that newly identified tasks have been taken into account.

The Appendix D quality plan is a slim document compared with plans commonly produced for petrochemical or nuclear projects. Many of these may run to between fifty and one hundred pages of subject matter specific to the project. Their issue is the fruit of weeks of discussion and negotiation between teams representing client and contractor. No doubt such documents have their value, but to impose procedures appropriate to major high-technology projects on to the typical average-size building site would be wasteful in the extreme. There is a need for a perspective which takes into account the type of work, the nature of the likely defects and the requirements of the market. The cost of preparing a quality plan must not exceed the likely benefits which can be expected to accrue from its use.

WORK INSTRUCTIONS

Section 5 of the sample quality plan identifies the elements of work which are to be covered by written work instructions. This is a topic which was discussed at some length in Chapter 5 in the context of the documentation requirements of quality standards. It was pointed out that work instructions on a construction site may come in a variety of forms and can originate from many different sources. Examples of these sources found in the sample quality plan are:

Project staff
Sub-contractors
Equipment or component manufacturers
Outside experts.

Some organizations publish standard instructions covering particular construction operations. These have their merits, particularly in the case of specialist operations which are wholly under the organization's own control. Typical examples of such operations may include the erection of timber frames for houses or the splicing of steel reinforcement using proprietory couplers. In such cases the work instruction becomes an extension of the technical specification and can have similar mandatory status.

When tackling its quality system documentation, Alias Construction considered whether it should prepare standard work instructions for the more common construction trades: earthworks, concrete, masonry, steelwork and so on. It came to the conclusion that the standard techniques used for such operations are well known to qualified tradesmen and that mandatory instructions emanating from head office would be superfluous, resented and probably ignored. The problems and defects which occur are seldom due to lack of knowledge of basic techniques, they are more likely to arise because well-known precepts of good practice have been ignored, or because particular designs require non-standard methods of work. Such problems are not susceptible to cure by prescriptive instruction documents. What is needed is good site discipline, well-trained supervisors and a procedure which causes potential problems to be identified and resolved in good time. These are the prerogative of site management, not head office management.

Having decided against issuing standard work instructions for the basic trades, Alias Construction became aware that site staff faced with the task of producing work instructions needed a ready source of information on available construction techniques. There was a risk that the same problem would be solved in different ways on different projects and that the lessons learnt on one site would not be passed to others. They therefore prepared a technical handbook containing a distillation of the company's technical experience which could be used as a basis of site work instructions. The contents of the book had no company-wide mandatory status, but its recommendations could be given mandatory effect on any site by incorporation into a project work instruction.

Given the variety of sources of work instructions, Alias Construction decided that it would not be practical to insist that all should follow a common format. However, to provide instruction to their staff, they issued Company Standing Instruction CSI OPS-04 which laid down the rules for preparing work instructions on site. The controlled receipt, storage and issue of work instructions is covered in CSI OPS-09 'Documentation and Change Control'.

IDENTIFICATION AND TRACEABILITY

The process of planning requires that packages of work should be broken down into discrete elements so that the relationships between them can be analysed, and time and resources can be allocated for their accomplishment. Having identified the elements, the planning engineer will number or label them to establish a code which he himself will use in constructing his plans and which will enable the user of plans to understand his proposals.

As an example of such an identification system, consider the planning of a reinforced concrete structure. The engineer will first divide the whole into a number of separate pours or lifts. He will number these and assess their material demands and work content. The latter will form the basis of his construction plans and the lift or pour numbers will supply a means of identification, both at the planning stage and thereafter during construction and beyond. They are likely to appear for example on the following documents:

Construction programmes
Reinforcement schedules
Formwork drawings
Concrete order forms
Pre-pour check sheets
Concrete delivery tickets
Pouring records
Concrete cube test results
Payment certificates.

These documents, if properly completed, can provide a trail whereby one can trace the origins of the materials of each element in a structure and can verify the circumstances of each stage in its construction. This traceability can be of significant benefit later in the life of a structure, particularly if there are reasons to doubt its structural integrity. Engineers called upon to decide the future of buildings constructed with high-alumina cement or with aggregates susceptible to alkali–silica reaction, for example, will find such records invaluable.

INSPECTION AND TEST PLANS

Having established the sequence and method of construction, the next step is to determine the system whereby management ensures that the finished product will comply with the specification. Construction specifications are lengthy and detailed documents and they frequently contain cross-references to other specifications such as those produced by national standards organizations. It is often the case that careful study and research are necessary to establish the totality of inspections and tests which are necessary to prove compliance. Such studies are best carried out at the planning stage with the preparation of inspection and test plans.

The purpose of an inspection and test plan is to assemble in one document all the testing and inspection requirements relevant to a particular operation or element of work. The plan, which may be presented as a schedule or flow chart, will list and reference all the relevant tests and inspections in the sequence in which they should be performed, together with the documentation to be used to record the results. Figure 8.1 is an illustration of a typical inspection and test plan. It relates to concreting operations to be undertaken in the course of construction of the supermarket which is the subject of the quality plan reproduced in Appendix D. The form is pre-printed with columns in which the user may list the checks which have to be carried out, the references which define the acceptance criteria and the documents to be used to record verification. The two right-hand columns headed 'N' and 'H' are for use when the purchaser has stipulated that he requires to be given advance notice of the contractor's intentions. It is envisaged that these requirements may relate to one or other of the following:

Notification *points (N).* These are stages in the construction process of which the purchaser has to be made aware in advance in order that he may witness tests or approve actions which the contractor proposes to take. In the event that he elects not to exercise his right to be present or to give his approval, then provided the necessary notice has been given, his absence or lack of response may be taken to signify approval.

Hold *points (H).* These are points beyond which the purchaser has specified that work may not proceed without approval by designated individuals or organizations.

The example given in Figure 8.1 relates to conventional work where the inspection and test plan has been produced by the contractor in order to clarify the specification requirements for the benefit of his own staff. It would be up to the contractor to decide whether to offer it to the Resident Engineer or Architect for his approval. In nuclear works it is a specification requirement that: 'A performance test programme shall be established where necessary to ensure that all testing required to demonstrate that items will perform satisfactorily in service is identified, performed and documented' (BS 5882). The inspection and test plan then becomes a controlled document, subject to the approval of the purchaser and not to be amended without his agreement.

A	ARCADIA SUPERMARKET TUNBRIDGE WELLS	Document No: ASTW - ITP - 2
		Page: 1 of 1
		Date: June 1988

INSPECTION & TEST PLAN FOR: *Concrete*

	ACTIVITY	ACCEPTANCE CRITERIA	VERIFICATION DOCUMENT	CLIENT N	CLIENT H
1.	*Approval of Readymix Supplier*	*Specification Clause 5.3*	*QSRMC Certificate*	✓	
2.	*Approval of Reinforcement Supplier*	*Specification Clause 4.5*	*CARES Certificate*	✓	
3.	*Approval of Construction Joint Layout*	*Specification Clause 8.1*	*Drawing ASTW-T-101*		✓
4.	*Approval of Transportation Placing and compaction*	*Specification Clause 13.1*	*Work Instruction No.2*		✓
5.	*Setting-out*	*Drawings & Specification Clause 3.0*			
6.	*Reinforcement*	*Drawings & Specification Clause 9.0*			
7.	*Embedded Items*	*Drawings*	*Pre-placement Check Form AC - 263*		✓
8.	*Formwork*	*Drawings & Specification Clause 6.0*			
9.	*Clean-up*	*Specification Clause 9.0*			
10.	*Approval of Exposed Aggregate Finish*	*Specification Clause 7.0*	*Trial Panel*		✓
11.	*Pour Concrete*	*Specification Clause 3.12*	*Placement Record AC - 264*	✓	
12.	*Curing*	*Specification Clause 3.14*	*Curing Record AC - 265*		
13.	*Cube Tests*	*Specification Clause 18.0*	*Cube test Results AC - 189*	✓	

N : Notification Point **H : Hold Point**

Signed for Sub-Contractor	*P Murphy*	**Date**	*4 July 1988*
Signed for Main Contractor	*A J Watling*	**Date**	*5 July 1988*
Signed for Purchaser		**Date**	

Figure 8.1 Inspection and test plan.

QUERIES

One of the principal benefits of the planning process is that it exposes any errors, gaps or discrepancies which may exist in the drawings and specifications defining the purchaser's requirements. To resolve or clarify such matters, it is useful to establish a documented procedure whereby the contractor's queries and the designer's responses can be recorded formally. The main contractor is usually the party in the best position to institute such a system. Not only will his own planners require responses to queries, he will also be in receipt of similar enquiries from his sub-contractors and suppliers. All will benefit if a standard method of documentation is adopted. Figure 8.2 illustrates a typical 'Technical Query' form. This identifies the originator, date and detail of the query and provides space for the response. It supplies a record of what has transpired and facilitates follow-up at review meetings.

Control

RECEIPT OF MATERIALS ON SITE

Verification of compliance with specifications commences when materials or supplies are received at the site. The criteria for compliance are usually defined in the specifications, together with the appropriate sampling and testing regimes.

Testing and inspections are not the only acceptable means of verification of compliance. Materials which carry product certification such as the BSI Kitemark or the British Board of Agrément Certificate should not need to be tested on receipt, nor should those accompanied by other evidence of control at source such as a compliance certificate issued by a supplier of proven capability. It also needs to be remembered that random sampling and spot checks are a satisfactory form of verification. There is therefore no need always to check every brick in a consignment or to open every pot of paint to check that it is the right colour. The extent to which materials should be inspected on receipt should be carefully planned, bearing in mind the costs of inspection and the possible penalties which may arise from building sub-standard items into the works.

Procedures should ensure that non-conforming materials are quarantined or otherwise prevented from being used.

SPECIAL PROCESSES

In construction work, it is seldom possible to establish compliance with specification solely by examining the finished article. The final standards

A		QUERY No.
TECHNICAL QUERY		

Query addressed to:

Type of Work:

Location in Works:

QUERY:

Drawings/Specification references:

Query raised by: .... Job Title: .... Date: ....

Department/Company: ....

ANSWER:

Signed: .... Job Title: .... Date: ....

Department/Company: ....

Distribution:

Figure 8.2 Technical query.

achieved are influenced by many intermediate factors such as the quality of the raw materials, the method of processing, the levels of discipline of the work force and so on, and each of these needs to be controlled if the final results are to be satisfactory. In the jargon of quality system standards, such operations are termed 'special processes', a term defined in BS 5750 as:

'processes, the results of which cannot be fully verified by subsequent inspection and testing of the product and where, for example, processing deficiency may become apparent only after the product is in use'.

Examples of special processes carried out on a construction site include the placement and compaction of concrete, welding, tunnel grouting and painting. All these operations depend on operator skill and care. The standard stipulates that to ensure that specified standards are met, such processes require 'continuous monitoring and/or compliance with documented procedures'. The word 'monitor' is frequently used in the various parts of BS 5750, but is not included in BS 4778 *Quality Vocabulary*. It does, however, appear in the following definition of 'surveillance' which is extracted from BS 5882 *A total quality assurance programme for nuclear installations*:

Surveillance Monitoring or observation to verify whether an item or activity conforms to specified requirements.

This is a long-winded way of describing the kind of supervision normally undertaken on a site. It confirms that concepts of quality assurance are not a substitute for the conventional supervision of work. Indeed, they support and demand its application. On the other hand, there are cases when monitoring or supervision are not by themselves a practicable means of ensuring compliance with specification. An example of such a case is welding. This is a highly skilled operation, almost entirely dependent upon operator technique. Continuous monitoring of welders by a supervisor would not only be wasteful, it would be ineffective since only the welders themselves are in a position to see and judge the quality of their work. It is for such cases that BS 5750 recommends 'compliance with documented procedures' as an addition to, or substitute for, continuous monitoring.

The standards therefore offer a wide degree of choice on the techniques to be used for the control of special processes. At one end of the spectrum there are operations, for example the compaction of concrete, for which documented procedures would be of little value and which are best dealt with by traditional supervision by trained and experienced foremen or gangers. In contrast, operations like welding require that work be carried out by trained operatives subject to regular qualification and supervision and following approved documented procedures. The documented proceures required to control special processes have been discussed earlier in this chapter under the headings 'Work instructions' and 'Inspection and test plans'. Note that the standards require documented work instructions only 'where the absence of such instructions would adversely affect quality'. This discretion is best exercised during the preparation of the quality plan when the activities requiring written work instructions can be identified and resources allocated for their production.

DOCUMENT CONTROL

The subject of documentation tends to dominate the popular image of quality management. It is seen as a creator of paperwork, much of which is unnecessary and all of which costs money to create, distribute and store. This is unfortunate and need not be true. In a properly managed quality system all forms of documentation are reviewed regularly by responsible managers, and if found not to be serving a useful purpose, are scrapped. Nevertheless, much will inevitably remain and this must be controlled.

Chapter 5 identified two distinct types of document: those produced in advance of work (instructions of one kind or another) and those generated after work is done (records). Let us concentrate on documents in the former category which are used on construction sites. For these, a procedure must be established which will ensure that the right document, and only the right document, is available at the time and place it is needed.

The procedure should first define the documents which can directly influence the quality of the product. Typically, these may include the following:

Drawings
Specifications
Quality plans
Project procedures
Work instructions
Inspection and test plans.

Like company quality manuals and company standing instructions (see p. 79), these should be classified as 'controlled documents'. This means that a register should be kept of each document listing the date of origin or receipt of each version (or revision), and the dates and locations of issues. The procedure should identify the person responsible for holding and maintaining the registers and should define the system and documentation to be used.

To comply with BS 5750, the procedure for document issue has to be capable of ensuring that:

'(a) The pertinent issues of appropriate documents are available at all locations where operations essential to the effective functioning of the quality system are performed.

(b) Obsolete documents are promptly removed from all points of issue or use'.

These requirements may be satisfied by the use of a numbered document issue slip which identifies the recipient, lists the accompanying documents and has a tear-off section to be signed by the recipient as a receipt. This should be returned to the drawing registry together with superseded versions of the documents listed.

SAMPLING

Control and verification of compliance with specification are provided by inspection and testing. But however hard we may try, certainty of compliance with specification will always be beyond our grasp. Even if our quality plans were to require 100 per cent inspection and testing of every component, it would not assure perfection since there is always the possibility of faulty testing equipment or neglectful personnel. In many cases, even if it were desirable, 100 per cent testing would not be feasible — one can hardly test every match in a box to make sure it will light. In such cases, the practical solution is to take representative samples of a batch or lot and to judge the acceptability of the whole according to the performance of the sample. Such practices introduce risks which cannot be entirely eliminated. However, they can be reduced to an acceptable level by applying sound statistical theory to the sampling process. There are many British Standards and other publications which provide detailed guidance on such matters as the proper relationship of sample size to batch size, methods of sampling and the quantitative analysis of results.

It is possible for an entire quality system to be undermined by faulty sampling procedures. Records can be meaningless and controls can become ineffective. It is essential that procedures for sampling receive the close attention of responsible management and that clear and precise instructions are given to those undertaking the work.

INSPECTION, MEASURING AND TEST EQUIPMENT

The degree of confidence that may be had in measuring and testing procedures is dependent upon the accuracy and reliability of the measuring and testing equipment. Control must therefore be exerted to ensure that only the correct equipment is used and that it is systematically maintained and calibrated. A quality system should include a procedure which identifies the types of equipment needing calibration and prescribes the methods to be used, the frequency and the records to be kept.

BS 5750:Parts 1 and 2 commit the supplier to a lengthy list of actions to assure the accuracy of measuring equipment. These are no doubt essential measures in establishments where articles are being manufactured to fine tolerances, but they go far beyond normal custom and practice in the construction industry. If the requirements of the standard were to be interpreted literally and applied to all measuring equipment on a construction site, there can be little doubt that costs would rise substantially with only minimal benefits. For example, there is a requirement that the supplier shall 'calibrate . . . all inspection, measuring and test equipment and devices that can affect product quality at prescribed intervals, or prior to use, against certified equipment having a known valid relationship to nationally recognized standards'. To apply this rule to a builder's level, a pocket tape or similar site measuring equipment would not be reasonable.

On the other hand, there can be no denying that site measuring equipment does suffer wear and tear, and expensive errors can and do result from the use of equipment which is out of calibration. Good survey technique can minimize these effects and at the same time give warning of incipient instrument defects. For example, the reading of theodolite angles on both faces of the instrument will both reveal and compensate for most of the errors to which these instruments are prone.

Equipment used for linear measurement is particularly vulnerable to wear and damage, and subsequent errors cannot always be detected. On a large site or in a regional headquarters where instrument stocks are held, it makes good sense to establish a standard baseline. This should ideally be 100 m long with intermediate marks at 10 m intervals, and it should be measured using a steel band which has been calibrated against a national standard. The baseline can be used for routine checks of distance measuring equipment including tapes and electromagnetic instruments (EDM). Tapes which are found to be outside permitted tolerances should be scrapped and EDMs similarly defective should be returned to the manufacturer or his service agent for repair and adjustment.

Table 8.1 is offered as a notional schedule of measuring and testing equipment on a typical construction site. The frequencies of calibration suggested are minima and any equipment which suffers damage or is found to be measuring incorrectly should immediately be quarantined until checked and, if necessary, repaired and re-adjusted. In the case of a large project, a schedule of this type should be included in the project quality plan together with the nomination of the person to be responsible for ensuring that the calibrations are carried out and for keeping appropriate records.

Table 8.1 Calibration of equipment

Item	Minimum frequency of calibration	Calibration method
Survey equipment		
Theodolite	1 month	Standard theodolite checks
Level	1 week	Two-peg test
Laser level	1 week	Two-peg test
Survey tape	1 month	Check against baseline
EDM	1 month	Check against baseline
Weighing equipment		
Concrete batcher	1 month	Check against standard weights
Weighbridge	3 months	Calibration by supplier
Laboratory equipment		
Thermometers	6 months	Check against laboratory standard
Scales	6 months	Check against standard weights
Cube crushing equipment	1 week	Check against laboratory standard

INSPECTION AND TESTING

During the course of the work the checks and tests scheduled in the Inspection and Test Plans are carried out in order to control quality. Under conventional contractual arrangements, the primary responsibility for quality control rests with the contractor or supplier, with the purchaser or his representative undertaking a monitoring role. Such arrangements are consistent with quality system standards, although the latter impose requirements for the organizational status of those who carry out inspection and testing which are absent in normal construction conditions of contract. The purpose of these is to safeguard the independence and integrity of inspectors and checkers and to protect them from unreasonable pressures.

BS 5750:Parts 1 and 2 require that 'the responsibility, authority and the interrelation of all personnel who manage, perform and verify work affecting quality shall be defined'. This is stated to be particularly necessary for personnel who need the 'organizational freedom and authority' to carry out functions such as identifying and recording quality problems, verifying the implementation of solutions and controlling non-conforming products. They further require that people carrying out verification activities should be properly trained and that design reviews and system audits should be by persons independent of those responsible for the execution of work. These requirements are not incompatible with traditional construction arrangements. It is generally accepted that designers, contractors and sub-contractors should employ competent people who will check their own work as they proceed. A contractor is required to provide competent superintendence of the site and to implement a regime of control which will make sure that required standards of materials and workmanship are achieved. Likewise, sub-contractors are usually required to supervise and take responsibility for the quality of their work.

The question arises, however: Do quality control staff as conventionally deployed on construction work in fact have sufficient 'organizational freedom and authority' to discharge their responsibilities without undue pressure? Should there not be on each site, an inspection team which is separate from the engineers, sub-agents and foremen in charge of getting work done, and which reports to a higher level of management? This, it may be argued, is what is meant by 'organizational freedom and authority' and only such an arrangement will satisfy the standard.

In nuclear work, there is a clear requirement that suppliers should maintain separate inspection teams. BS 5882 *A total quality assurance programme for nuclear installations*, requires that the organizational structure and functional responsibility assignments shall be such that:

'(a) attainment of quality objectives is accomplished by those who have been assigned responsibility for performing work (e.g. the designer, the welder, or the nuclear facility operator); this may include interim examinations, checks, and inspections of the work by the individual performing the work;
(b) verification of conformance to established quality requirements is accomplished by those who do not have direct responsibility for performing the work (e.g. the design reviewer, the checker, the inspector, or the tester)'.

No doubt, the particular hazards of nuclear work justify this more rigorous approach. A common response of suppliers to BS 5882 is to appoint a single manager with responsibilities for both quality assurance and quality control. This 'QA/QC' arrangement is also frequently encountered in petro-chemical and off-shore work (see p. 56). To what extent is it reasonable to apply these principles to more mundane construction work?

The answer to this question should depend on the contractual arrangements between purchaser and supplier and the degree of supervision exercised by the purchaser himself or his representatives. The minimum need is for one party to carry out the tests and checks required for control purposes, and for another to provide independent verification (by monitoring and spot checks) that the tests and checks have been properly executed and truthfully recorded. Conventionally, the former role belongs with the contractor and the latter with the purchaser or his representative. A purist might respond that monitoring by the purchaser's representative does not count because it does not relieve the contractor of his responsibilities. On the other hand there has to be a practical limit to the number of tiers of inspection and verification. Only on very large or particularly sensitive projects can it really be cost-effective to require both QC and QA (or QA/QC) from the contractor and another level of QA from the purchaser's representative.

The wider implications of invoking the requirements of quality system standards in parallel with conventional conditions of contract are discussed further in Chapter 10. For most construction projects, the presence on site of a purchaser's representative with powers to carry out such surveillance (including spot checks) as may be necessary will provide an adequate degree of assurance. In such circumstances, the contractor's quality system, and any additional surveillance which it may introduce, will be at the contractor's discretion and not part of the contract.

As an alternative to the above, the purchaser may place the whole responsibility for control and assurance on the contractor, and restrict the role of his site representative to the approval of quality plans followed by system audits. To make such an arrangement effective it would be necessary for contractors to undergo a rigorous system appraisal as part of the pre-qualification process, and the contract documents would have to contain precise instructions specifying the structure of the contractor's quality organization, including the independence of those responsible for inspection and testing.

INSPECTION AND TEST STATUS

A system should be operated to indicate the inspection status of material during intermediate stages of processing or manufacture so that it is at all times possible to distinguish that which has been inspected and found acceptable from that which has not. In many cases, inspection status can conveniently be indicated by the attachment of stickers or labels. Another method is the use of a check list. For example, airliner crews use pre-printed check sheets to tick off all the checks they are obliged to carry out at each stage of a flight. When signed by the captain the sheets also provide verification that the airline's quality system is being complied with. Although the presence of a signature or initials against an item on a check sheet does not by itself prove that the item is correct, the necessity to make one's mark does have a deterrent effect on those who may otherwise turn a blind eye.

An example of a check sheet for use on a construction site is shown on Figure 8.3. This is one of the 'verification documents' scheduled on the Inspection and Test Plan illustrated in Figure 8.1. It serves to record that the checks specified to be made prior to placing concrete have been correctly

A	SITE :	CONTRACT No. :

CONCRETE PRE-PLACEMENT CHECK AND INSPECTION	POUR No. :

CONTRACTOR :	CONCRETE GRADE :	
LOCATION :	ADDITIVE :	
	TARGET SLUMP (mm) :	VOL. (m^3) :
	METHOD OF PLACEMENT :	
DRAWING No. :	METHOD OF CURING :	

APPROVAL IS REQUESTED TO PLACE CONCRETE ON :

.... / .... / 19........ at .... hrs.

CONTRACTORS SIGNATURE : ....

DATE : .... / .... / 19........

NB. 24 HOURS SHOULD BE GIVEN PRIOR TO CONCRETING

CHECKS		READY FOR INSPECTION		INSPECTION	DATE	REMARKS
		CONT. SIGN.	DATE			
SETTING-OUT						
REINFORCEMENT CONTENT						
REINFORCEMENT COVER						
JOINT DETAILS						
WATER BARS						
CLEANLINESS						
FORMWORK & TIES						
BUILT IN ITEMS	CIVIL					
	M & E					

NOTES :

YOU ARE CLEARED TO PLACE THIS CONCRETE

FOR CLIENT ....

DATE .... TIME ....

Figure 8.3 Concrete pre-placement check sheet.

carried out by the contractor's supervisor and the purchaser's inspector, that all was found to be in order, and that concreting may proceed. Record sheets of this kind provide a discipline for the checking of work. This discipline can be strengthened if they are also used to supplement or even supplant the quantity surveyor's measure for payment. A sub-contractor who knows he will not be paid until documentary evidence of compliance with specification is complete will take more care than one who has only to satisfy a quantitative measure.

For small and movable articles, it is often convenient to indicate inspection status by location. For example, in the receiving area of a store, specific areas may be marked out for goods received but not inspected, goods inspected and found acceptable, and a special quarantine area for goods found to be defective. Such areas may be partitioned and fitted with security locks to ensure that there is no possibility of sub-standard materials entering the manufacturing process.

CONTROL OF NON-CONFORMING PRODUCT

Despite the emphasis of our quality system on matters of prevention, it is inevitable that sooner or later the production or construction process will yield some defective work. When this happens it is necessary that there should be a procedure to prevent any further harm being done and to put right that which is wrong. There are three stages to be dealt with in this procedure:

Identification
Segregation
Disposition.

Identification is necessary to make sure that defective goods are not inadvertently mixed with conforming materials and put back into the production process or shipped to the customer. Methods of identification must be such as to prevent accidental removal, and may include marking with colour-coded paint, or fixing labels or tags. After marking, the non-conforming items should, if possible, be removed to special holding areas or quarantine stores until their disposition is decided. At this point it is necessary to establish who will make the decision. Obviously it must be someone with authority and with the knowledge to understand the implications of the decisions he is making. It may be necessary also to consult the purchaser or his representative.

If a non-conformance is significant and not capable of rectification, there will be no alternative other than to scrap the offending article. On the other hand, the purchaser may be prepared to accept it in its existing state, possibly for some less demanding purpose than that originally intended and perhaps at a reduced price. More likely it will be necessary to 'repair' or 'rework' the article. These are jargon terms, best defined in BS 5882, *A total quality assurance programme for nuclear installations*:

Repair The process of restoring a non-conforming characteristic to an acceptable condition even though the item may still not conform to the original requirement.

Rework The process by which an item is made to conform to the original requirement by completion or correction.

Decisions on acceptance, repair or rework need to be recorded, and it is advisable for suppliers or contractors to have standard forms for recording descriptions of non-conformances, the causes, concessions agreed to by the purchaser and details of any repairs or rework carried out. Repaired or reworked items must be re-inspected in accordance with the original quality plan before being fed back into the system or being handed over to the customer.

CORRECTIVE ACTION

'Corrective action' is a jargon term for measures taken by management to ensure that conditions which may impede or prevent the achievement of specification requirements are identified and corrected. The need for corrective action may be indicated by evidence that non-conformances have already occurred, for example through inspectors' reports or customer complaints. Alternatively, management reviews or audits may have revealed defects in the quality system which are likely to lead to non-conformances in the future. In either case, it is necessary to hold an investigation to establish causes and to decide on the actions necessary to eliminate or minimize the subsequent incidence of defective work.

The responsibility and authority for instituting corrective action should be defined as part of the quality system. In a factory or on a large construction site the role might be assigned to the quality assurance manager. However, while the quality assurance manager may be able to co-ordinate, record and monitor corrective actions, the analysis and execution must involve the line management responsible for the work or product in question. On smaller construction sites it will normally be the project manager or agent who will initiate corrective action.

A useful first objective for such an investigation is to determine whether or not the circumstances which have led or may lead to defective work are susceptible to correction by the operatives alone, or if action by management to change the system is required. To enable them to function properly, operatives require:

Knowledge of what they are supposed to do
Knowledge of what they are in fact doing
A means of controlling what they are doing.

If these needs have been met, then any defective work which may arise can reasonably be attributed to lapses by the operatives. If, however, it becomes evident that operatives are not properly equipped, trained or informed, then management action to put this right is needed. It is important to distinguish between these two cases. When people make mistakes, it is necessary that they should be corrected, but they should not be blamed for errors over which they have no control. Inadequacies of equipment, training or information are symptoms of a faulty quality system. Only management can change the system, and unless they do so, defective work will continue to occur.

The nature of construction work is such that a large proportion of non-conformances are the results of lapses by operatives. A survey published in 1985 by the United Kingdom Building Research Establishment entitled *Achieving Quality on Building Sites* showed that of nearly 200 examples of defective work attributable to site management (as opposed to design management), on some 50 construction sites, about 45% were brought about by lack of care, nearly twice as many as were caused by lack of skill or knowledge. The only corrective actions likely to cure lack of care are increased supervision or the reduction of financial incentives to maximize output. The trouble is that whereas such actions may well be beneficial in the long term, their immediate effect is to increase a contractor's costs and preclude him from competing successfully for work. The market therefore tolerates a level of defective work regarded by many as unacceptably high, but which is due in large measure to the purchasers' own procurement policies.

Why should this be so? Perhaps the answer lies in the relationship between the degree of perfection which the purchaser would like and that which he is prepared to pay for. If there is a risk of one in a thousand that a particular defect will occur, an action to prevent it will be unnecessary 999 times in every thousand. In construction work it is often cheaper to repair or tolerate the one defect when it occurs rather than take unnecessary preventive measures 999 times. This argument will be of little comfort to the one purchaser in a thousand who suffers the defect, but in the long term it may well be in the interests of the thousand to bear this risk rather than pay the cost of perfection. This argument can only be taken so far. Defects caused by lack of care seldom follow a predictable statistical pattern, and this makes their effects particularly difficult to evaluate and to prevent.

Verification

The final phase of a site quality management system is the provision of objective evidence that work has been carried out in compliance with the drawings and specifications. Quality system standards envisage that most of this evidence will be provided by documentary records, as was discussed in Chapter 5. They require that records should be maintained both to provide proof of compliance and to demonstrate the effectiveness of the quality system.

Traditional contractual arrangements in the construction industry vary in their demands for the provision of documentary evidence. The JCT Standard Form empowers the Architect/Supervising Officer to require the contractor to provide vouchers to prove that the materials and goods he is supplying comply with specification. The ICE Conditions of Contract do not contain such a requirement, although specific instructions to this effect are frequently included. These arrangements are compatible with BS 5750 which states: 'Where agreed contractually, quality records shall be made available for evaluation by the purchaser or his representative for an agreed period'.

On construction sites, additional verification for the purchaser's benefit may be provided by the activities of his professional advisers and representatives. These conventionally include the Engineer and his Representative on a civil engineering site and the Architect/Supervising Officer on a building project. When work is carried out under a management contract a similar verifying role is undertaken by the management contractor. All these parties normally sign certificates of completion of work for payment purposes, and so must assure themselves that the work measured for payment is in compliance with specification. The means by which they gain this assurance, and the responsibilities which they thereby assume, depend upon their terms of engagement. These matters have already been discussed at some length in Chapter 2, and the theme will be developed further in Chapter 10.

Finally, a quality system provides verification and assurance through the practice of reviews and audits, which takes us to the next chapter.

9

QUALITY AUDITS

The background to quality auditing has already been discussed in Chapters 4 and 7. This chapter will deal with the techniques of auditing as they may be practised in the construction industry.

Principles of auditing

All management decisions should be based on a knowledge of the relevant facts. Decisions made without a knowledge of the facts, or those based on false information, are likely to be bad decisions which will not achieve their desired objectives.

In most organizations there is no shortage of information. People are bombarded with paper, reports, memoranda and talk. The difficulty is to select the information which is relevant, and which can be relied upon to be factual. In this respect those at the bottom of the management tree are often more fortunate than those at the top. They need do no more than obey properly authorized instructions and ignore everything else. The senior manager, however, is in a more difficult position. Much of the information he receives will have flowed upwards through the various tiers of management in the hierarchy, and in the process it is likely to have become distorted. His subordinates will try to ensure he hears only what they want him to hear. The bad news may be filtered out or so submerged in extraneous material that it becomes impossible to find. This accounts for many of the delusions held by some managers who believe that all is well because no-one tells them that it is not. It is only when disaster strikes that they become aware of the true facts and by then it may be too late.

It is for these reasons that finance managers require that audits should be held by trained and independent experts to provide information which can be trusted. Financial audits are carried out in accordance with formal procedures to provide specific facts which will enable sound commercial judgements to be made. Quality audits fulfil a similar purpose to these financial audits. Indeed, they are complementary. Financial assurance will provide a more complete view of a business when accompanied by quality assurance. If profits have been achieved at the expense of quality, they are likely to prove illusory. Companies seeking long-term survival need reliable information on the quality of their product as well as on their current financial status.

There are other ways in which quality audits can contribute to the health of an organization. The lines of communication which they provide can enable people at all levels to contribute to the improvement of management systems. People in the lower levels of a hierarchy are often in a better position to judge the effectiveness of systems than are their seniors. They know when procedures have become out of date or superfluous, and will say so if asked. Quality audits provide opportunities for them to make suggestions about the activities which affect their working lives. This can bring significant benefits in terms of morale and motivation.

So far we have discussed reasons for self-auditing. Many organizations audit also their suppliers or contractors. This may be because they want information on their capabilities to enable them to decide whether to enter into a contract or otherwise do business, or they may already have a contract and wish to satisfy themselves that the supplier's quality system is continuing to function properly to assure the quality of their purchase.

Sceptics, and there are many of these in the construction industry, may feel inclined at this stage to point out that there is nothing novel in the concept of managers talking to people in the lower echelons and seeking their advice, nor is it anything other than common practice for purchasers to investigate, pre-qualify and oversee their contractors and materials suppliers. Unfortunately, while such informal arrangements may serve a limited purpose, in large organizations they cannot be relied upon to provide all the information needed. To pursue the analogy with financial audits, few finance managers would be prepared to judge commercial viability on the basis of casual conversations. Like financial audits, quality audits have to be organized on a formal and official basis if they are to achieve their objectives. These and other characteristics are summed up in the standard (BS 4778) definition of a quality audit:

Quality audit A systematic and independent examination to determine whether quality activities and related results comply with planned arrangements and whether these arrangements are implemented effectively and are suitable to achieve objectives.

This definition identifies that the purpose of a quality audit is to compare 'quality activities and related results' with 'planned arrangements'. In many organizations the planned arrangements consist of documented procedures put together in accordance with a system standard such as BS 5750. When this is so, the first stage of an audit consists of checking whether or not the documented system does, in fact, comply with the standard. This is then followed by a second stage at which it is established whether or not the procedures are being implemented effectively. Audits carried out against a system standard have the advantage that their results can be related to an agreed benchmark. The standards also provide a common format to which both auditors and auditees can become accustomed. On the other hand, although it may be desirable, compliance with a standard is not essential. Any organization which has defined its objectives and has established planned arrangements to achieve these objectives can be audited. It follows, however, that in the absence of planned arrangements no comparison can be made and auditing as defined cannot take place.

Audits which examine quality systems only are also known as 'assessments'. Those which go further and examine the product as well are known as 'product conformity audits'. This chapter will deal only with audits of systems, since the checking of product conformity by inspectors and clerks of works is already a well-established routine within the construction industry and does not require further discussion. Hereafter, the terms 'quality audit' and 'quality assessment' will be synonymous.

Types of audit

Most audits fall into one of three categories.

INTERNAL AUDITS (1ST PARTY)

These are audits undertaken by an organization to examine its own systems and procedures. They may be performed by people from within the organization or by teams hired from outside. Note that internal auditors must be independent of the people being audited in order that their objectivity will not be compromised.

EXTERNAL AUDITS (2ND PARTY)

These are audits undertaken by an organization to examine the quality systems of its suppliers. Again, they may be performed by in-house teams or by outside agencies hired by purchasers.

THIRD-PARTY AUDITS

These are audits undertaken by bodies with no existing or intended contractual relationship with either the purchaser or the supplier. They are almost invariably made against a recognized system standard and include audits by accredited certification bodies (see p. 111).

Auditor selection and training

Quality auditing is a skilled function and, like all other skills, requires specific personal talents, specialized training and broad experience. Let us consider these requirements in turn.

PERSONALITY

In many respects, quality auditing is akin to police work. It involves the same careful sifting of evidence and the questioning of witnesses to arrive at the truth. Cross-examination can arouse emotions between questioners and respondents, with aggression on one side and apprehension and resentment on the other. The relationships are inevitably tense and if not properly handled, can quickly deteriorate into personal antagonism. An auditor must, therefore, have a good understanding of human behavioural patterns and be able to control the emotional temperature if there are signs of overheating; he or she must be tactful and diplomatic. In addition to these qualities, auditors must have orderly minds capable of sorting and analysing the evidence presented to them. They must be patient, firm, and good listeners. Above all, they must be honest and trustworthy — if not, their findings will always be suspect.

TRAINING

Auditors must be trained in the techniques of auditing. These include audit preparation, the tactics of auditing and the preparation of audit reports. To be able to audit against quality system standards, they must have a thorough knowledge of the contents of the standards and the ways they may be interpreted in the context of the particular industry sector concerned. Training courses covering system standards and the techniques of auditing are available from many sources. Most last for about one week and include periods of tuition and case-study work. In the United Kingdom, the Institute of Quality Assurance administers a scheme for the approval of auditor training courses and for the certification of trained auditors. This is known as the Registration Scheme for Lead Assessors and Assessors of Quality

Assurance Management Systems. To become registered as an Assessor, a candidate has to hold certain educational qualifications and to have successfully completed a recognized training course. Assessors can achieve Lead Assessor status by participating in a minimum number of assessments under controlled conditions.

EXPERIENCE

Auditor training is essential but experience, too, is needed. This includes practical experience of auditing and experience of the particular type of work being examined. In the opinion of the author, it is a fallacy to suppose that auditor training and auditing experience can by themselves enable a person to examine and make judgements on work of which he has no knowledge or understanding. Without the latter, it will be all too easy for an auditee to exploit his ignorance. Furthermore, there is a risk that any adverse reports he makes will generate ill-feeling. Most people are prepared to be judged by their peers, but are likely to take offence if criticized by those whom they consider to be inferior in knowledge or experience. In this respect, the analogy with financial auditing breaks down. Financial systems do not vary much from one industry to another, and a good accountant can satisfactorily audit all kinds of commercial operations. But quality systems do vary, and quality auditors need practical experience of the work in hand if they are to come to sensible judgements.

The need for relevant experience has to be taken into account when selecting audit teams. Some organizations maintain a permanent staff of qualified assessors who carry out all internal and external audits. This arrangement has the advantage that the people concerned are able to develop their auditing skills to a high level, but they are unlikely to have experience of all the different types of work they will be called upon to audit. An alternative is to maintain a small core of experienced lead assessors and supplement them with other line managers who have also received training in the techniques of assessment. The advantages of this arrangement include the following:

1. Audit teams can benefit from the practical experience of line managers.
2. Line managers have an opportunity to examine outside bodies and other parts of their own organizations. This cross-fertilization can be a most valuable source of new ideas and understanding.
3. By auditing others, line management will learn how to respond to being audited themselves.

The right to audit

A successful audit requires active co-operation between auditor and auditee.

This co-operation will not automatically be forthcoming. There are many reasons why an auditee may feel resentment at being coerced into submitting to examination by outsiders, and this natural emotional response may make it quite impossible for even the most skilled auditor to carry out his task. It is the duty of those commissioning the audit to establish its legitimacy in the eyes of both parties.

In the case of external audits, the consent of the auditee may derive from the hope or expectation of future business, as in the case of pre-contract assessments. Alternatively, the right of the purchaser to audit may be written into a contract which, by definition, signifies consent between the parties that certain actions will be taken. Similar considerations will apply in the case of third-party audits.

Establishing the legitimacy of internal audits can be more difficult. Auditees have neither the motivation of benefits to come nor are they under a contractual obligation. It thus becomes the duty of the senior managers concerned to agree between themselves the ground-rules under which audits will be held. These must then be communicated to both auditors and auditees so that each side is aware of the other's rights and obligations. It is recommended that this communication should be in the form of a written brief signed by the managements of both parties.

Audit procedures

There is not, as yet, any national or international standard specifying the procedure to be followed in carrying out an audit or assessment. On the other hand, there does exist a reasonable consensus on the most suitable procedures to be followed, and this forms the basis of the training courses recognized by the Institute of Quality Assurance. The following paragraphs describe the principal steps in a conventional assessment. These are illustrated on the flow chart in Figure 9.1.

PREPARATION

On receipt of an instruction to carry out an audit, the lead auditor's first action should be to make contact with the auditee and establish that there is a common understanding of the background to, and purpose of, the audit. He should then agree dates for the supply of preliminary information, and for the audit itself. Having established the programme, the auditor should confirm it in writing and, at the same time, make a formal request for the information required for his preliminary studies. These studies are an essential part in the audit programme. They can ensure that the audit will proceed smoothly and that the assessors will be able to secure all the information they need within the shortest possible time with minimum disruption to the auditee's operations.

In carrying out his preparations, the auditor will seek to achieve the following objectives:

1. To establish whether the auditee has a documented system capable of satisfying the requirements of the audit.
2. To select matters in the documents which require clarification.

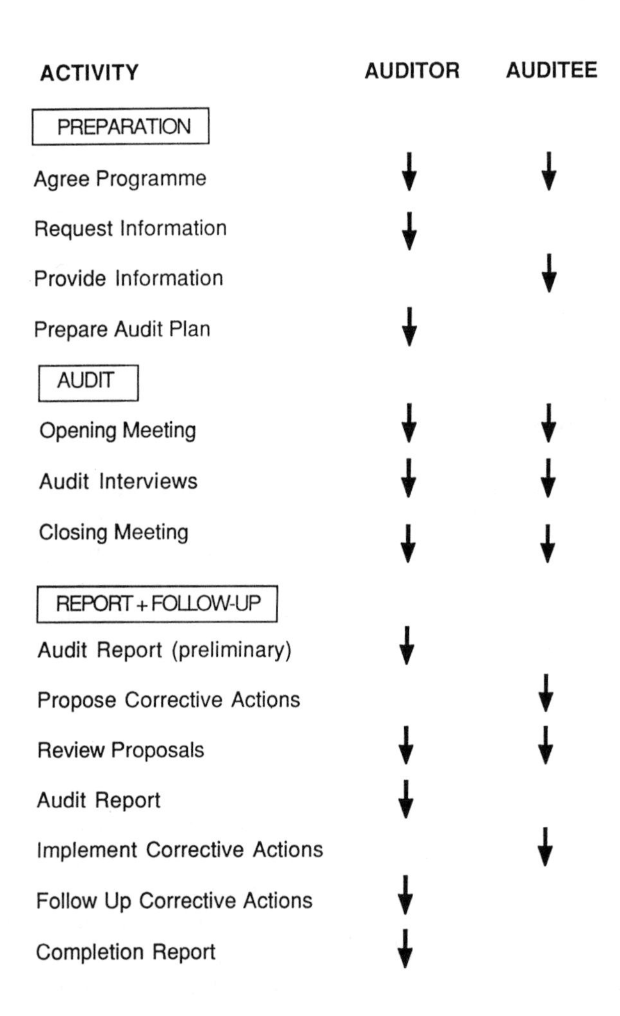

Figure 9.1 Audit flowchart.

3. To identify the functions or departments in the auditee's organization which he will wish to examine and to draw up a check-list of questions applicable to each.

In most cases, a copy of the appropriate company quality manual or project quality plan will provide the necessary information. On occasion, it may also be necessary to call for copies of procedures or standing instructions, but this is to be avoided if possible since procedures manuals tend to be bulky documents. Many auditees will resist requests to submit procedures manuals on the grounds of confidentiality — they are prepared to offer them to an auditor for inspection but will not permit them to be taken from their premises. This is a reasonable stance for an auditee to take, and it should be respected.

If the purpose of the audit is to determine whether the auditee's quality system complies with a standard such as BS 5750, the first step will be to examine the documentation to check that each relevant clause of the standard is addressed. As discussed earlier in this chapter, this is not the only basis on which a system can be assessed. The alternative is to establish whether or not the auditee's system satisfies the requirements of the party which has commissioned the audit. This demands that the auditee's declared objectives be acceptable and that the system should be capable of meeting these objectives. The question of objectives is important. An organization which states that it exists solely to maximize its own profit is not likely to endear itself to its customers as the profit will be gained at their expense. The exercise of examining system documentation to establish its compliance with a standard or its ability to achieve acceptable objectives is sometimes know as a 'system audit'.

Having completed his examination of the documents, the auditor should prepare an audit plan identifying the sequence and timing of the interviews he wishes to conduct. This should be sent to the auditee for his consideration with a request that the appropriate members of staff should be asked to hold themselves available at the stated times. A typical audit plan is illustrated in Figure 9.2. This is for a comparatively small-scale audit undertaken by a team of only two persons. With a larger team it is possible to divide the audit team into groups and conduct a number of interviews in parallel.

THE AUDIT

It is usual to commence the audit with an opening meeting attended by the audit team and the heads of the auditee's departments which are to be assessed. The purpose of this meeting is to ensure that all concerned understand the background and purpose of the audit and are aware of the procedure to be followed. Here is a typical agenda for an opening meeting:

	AUDIT PLAN	
Project:	*Crooktown Prison*	
Date:	*28-29 January 1988*	
Audit Type:	*Internal*	
Audit Criterion:	*Company quality system*	
Auditors:	*H. Pile, R.S. Jay*	

DATE & TIME	SUBJECT	PRESENT
28th Jan *09.00*	*Opening Meeting*	*Project Manager* *Chief Engineer* *Construction Manager* *Commercial Manager* *Quality Engineer* *Planning Engineer*
09.30	*Tour Site*	*Quality Engineer*
10.30	*Organization Structure*	*Project Manager* *Quality Engineer*
11.30	*Document Control*	*Chief Engineer* *Planning Engineer* *Quality Engineer*
14.00	*Materials Purchase, Sub-contractor selection and appointment*	*Commercial Manager* *Quantity Surveyor* *Buyer* *Quality Engineer*
16.00	*Material Receipt and Storage*	*Storeman* *Quality Engineer*
29th Jan *09.00*	*Control of work* *Sub-contractor control* *Control of non-conformance*	*Construction Manager* *Quantity Surveyor* *Chief Engineer* *Quality Engineer*
11.00	*Inspection and Testing*	*Chief Engineer* *Quality Engineer*
14.00	*Records*	*Chief Engineer* *Quality Engineer*
15.00	*Preparation of Audit Summary*	
16.00	*Closing Meeting*	*Project Manager* *Chief Engineer* *Construction Manager* *Commercial Manager* *Quality Engineer* *Planning Engineer*
17.00	*FINISH*	

Figure 9.2 Audit plan.

1. Introdution
2. Background and purpose
3. Confirmation of audit plan
4. Nomination of escort
5. Identification of auditee management representatives
6. Logistics: working hours, office accommodation, lunch, closing meeting, etc.
7. Questions.

The nomination of an escort is to ensure that there is someone to guide the audit team from one department to another and to introduce them to the management representatives who will be responsible for responding to questions.

After the opening meeting, it is recommended that the auditors should seek to obtain an overall perspective of the organization which they are assessing. If it is a construction project, a quick tour of the site is valuable. If it is an office, then it is useful to ask for a brief description of the various departments and the interfaces between them. This initial overview should not be allowed to take up too much time. The auditors must be constantly on their guard against time-wasting by the auditees. Lengthy presentations, liquid lunches, frequent coffee breaks and similar ploys may be used to consume time and distract the auditor's attention from areas which the auditee does not want probed too thoroughly.

The auditors may now move on to examine compliance with the documented procedures. Their purpose should be to seek 'objective evidence' that the work-force is aware of the requirements of the system, that they understand their personal roles and that the system is being put into practice. This requires that they ask questions and, when necessary, probe the truth of the answers they receive. Most of the enquiries will be directed to the appropriate management representative, but it will also, on occasion, be necessary to put questions to supervisors and operatives. As a matter of courtesy, the management representative's permission should be sought before so doing.

At the start of each interview, it is suggested that the auditors should resolve any queries which may have arisen from their study of the documentation. They should point out the deficiencies they have identified and invite the auditee to correct any misconceptions that may exist. The auditee may be able to offer further documentation which will provide a satisfactory resolution of the problems raised. It is at this stage of the audit that the auditor's skills in interviewing come to the fore. He should phrase his questions carefully, making them as open-ended as possible. Having asked a question, he should listen carefully to the response and make sure he has understood it. He should make notes as he goes along and for this purpose it is useful to frame question check-lists with spaces for responses alongside each question. Written notes may also be amplified by further comments recorded on a dictating machine.

When deficiencies are detected, these should be discussed at the time with the management representative to make sure that there is no misunderstanding. Some auditors make it a practice to prepare a 'deficiency report' for on-the-spot signature by the management representative each time a discrepancy against a procedure is observed. Others may regard this practice as unnessarily officious but it does have the undoubted benefit of preventing future arguments about the facts.

On completion of the interviews it is usual to hold a closing meeting at which the auditors present a verbal summary of their findings. Time should be allowed in the programme for the auditors to review the responses to their questions and to prepare their presentation. As a preface to the presentation, it is wise to point out that an audit is only a sample, that the absence of non-compliances in any particular area at the time of audit cannot be taken as proof that they do not exist at other times, but that if the sample did contain discrepancies or deficiencies then the likelihood exists that they would continue to occur unless prevented.

After the findings have been presented, the management representatives should be given any further explanations they may ask for, but it is wise not to engage in further debate about the facts or their interpretation. On the other hand, if the auditee is able to demonstrate that any particular audit finding is incorrect, it should be withdrawn without further argument and suitable apologies should be given. After completing their exposition, the auditors should outline the procedure to be followed for the issue of formal reports and for recording and implementing management's responses. They may then, if there are no further questions, thank their hosts for their hospitality and depart.

REPORT AND FOLLOW-UP

The audit reporting procedure will depend upon the type and purpose of the audit. Its primary aim will be to provide information to the party which has commissioned the audit. In the case of a pre-contract assessment, this is the only purpose. The purchaser requires information in order to determine his future relationship with the supplier. He may, at his discretion, inform the supplier of his findings, but is not under a duty to do so, since the rectification of deficiencies in the auditee's system is not his concern. On the other hand, in the case of an internal audit, the main objective is to improve the performance of the auditee and it is therefore necessary to inform him of the auditor's findings, to obtain his response and to follow up corrective actions to put right any deficiencies which may have emerged. Similar considerations will apply in the case of audits by third-party bodies to determine eligibility for registration, or when there is a contractual relationship between the parties which empowers the purchaser to oblige the supplier to carry out corrective actions to his approval. In such cases as these, the audit is not complete until corrective actions have been implemented and their implementation has been followed up and checked by the auditor.

A typical format for an internal audit report is illustrated in Figures 9.3 and 9.4. Figure 9.3 is a front sheet which contains important reference data and a summary of the principal findings. The front sheet is designed to be accompanied by a series of Corrective Action Requests (Figure 9.4), each of which deals with a specific audit finding and has spaces for entering the auditor's report, the auditee's response, the agreed corrective actions and, finally, verification that corrective actions have been implemented.

It may be possible for the audit findings and responses to be fully documented before the audit team leaves the auditee's premises. More often, the auditee will require time to consider his responses. When this is so, the auditor may issue the auditee with a preliminary report consisting of the front page with its summary of findings, together with a set of Corrective Action Requests with Part 1 completed for each finding. The auditee's management should be asked to return the latter with their responses and proposals for corrective actions entered in Part 2. There may then be a period of negotiation between auditor and auditee to agree the corrective actions and the programme for implementation. When agreement has been reached, the auditor completes Part 3 of the Corrective Action Request and the Audit Report is issued to the nominated list of recipients. When it is decided to adopt this staged procedure, it is essential that both the auditor and auditee adhere to a mutually agreed timetable for each step.

It is generally desirable, and in the case of internal audits it is essential, that decisions on corrective actions should rest with the auditee's management. The auditor may offer advice but he is unlikely to have, or want, the power to impose his own solutions. If he believes that an action proposed by the auditee is an inadequate response to a particular audit finding, or if the auditee fails to implement an agreed corrective action, his recourse is to the party who commissioned and established the legitimacy of the audit. In this respect, audits which are undertaken against a system standard are easier to conduct and to report upon than those which require assessment of the acceptability of the auditee's objectives and planned arrangements. In the former, findings can be specific and objective — either the standard is being complied with, or it is not. There is no call for subjective judgements on the suitability of procedures and documentation or on their commercial justification. On the other hand, it can be argued

<table>
<tr><td colspan="2">A   **AUDIT REPORT**</td></tr>
<tr><td colspan="2">AUDIT REF: AUDIT DATE: Page........ of ....</td></tr>
<tr><td>Auditee & Address

Telex No.
Telephone No.</td><td>Scope of supply or service:</td></tr>
<tr><td>Relevant Project or Purchase Order:</td><td rowspan="4">Auditee Representatives</td></tr>
<tr><td>Audit Criterion
Type of Audit:</td></tr>
<tr><td>Auditors:

Of:</td></tr>
<tr><td>Previous Audit Ref:
Previous Audit Date:</td></tr>
<tr><td colspan="2">AUDIT SUMMARY</td></tr>
<tr><td colspan="2">Distribution</td></tr>
<tr><td colspan="2">DATE .... SIGNED .... AUDITORS</td></tr>
</table>

Figure 9.3 Audit report.

A	AUDIT REPORT CORRECTIVE ACTION REQUEST	
AUDIT REF:	AUDIT DATE:	Page .... of ....
USE ATTACHED SHEETS FOR CONTINUATION AS REQUIRED		
PART 1. AUDIT FINDING No.		
Signed Auditor(s)		Date: ....
PART 2. AUDITEE MANAGEMENT RESPONSE		
Signed Auditee		Date: ....
PART 3. AGREED CORRECTIVE ACTIONS & COMPLETION DATES		
Signed Auditor(s)	/ Auditee	Date: ....
PART 4. FOLLOW-UP & CLOSE OUT		
Signed Auditor(s)		Date: ....

Figure 9.4 Corrective action request.

that an audit which takes no account of the commercial background in which an organization operates is only of limited value.

The final stage in the audit consists of the auditor checking that each of the corrective actions has taken place by the agreed date. He may then sign Part 4 of the appropriate report sheet and close his file.

The auditee's reponse

So much for the techniques of the auditor; what of the auditee? He too needs to make preparations and to train his staff so that they will give a good account of themselves and present their company or department to its best advantage. First impressions are important. For this reason it is vital that the documentation sent to the auditor in advance of the audit should be well-presented, clearly worded and indicative of competence and good order. It should give the auditor the information to which he is entitled in a simple and straightforward manner without unnecessary padding and verbosity. While presentation is important, few auditors are likely to be impressed by glossy binders or superfluous bulk. Clarity, brevity and above all, truth, are the qualities which will win their favour.

Similar precepts should rule during the audit interviews. Questions should be answered honestly and openly. Auditors quickly learn to sense when they are being deceived, and the simple request 'show me' can be a trap for those who might try to draw the wool over their eyes. It is wise to establish in advance who will respond to questions in respect of each subject or department. Some auditees ask for all questions to be put to a chairman who either answers them himself or calls upon one of his team to contribute. This may seem pedantic, but it has the advantage that it prevents members of the auditee's team arguing among themselves as to how a particular function is carried out. Such signs of internal dissension will be seized upon by an auditor as evidence of lack of control on the part of the auditee, and should be avoided at all costs.

Auditees should be aware that there is a positive side to audits and that there are potential advantages which they can gain from the process. A professionally executed audit can provide independent management consultancy without charge. It can goad management into making improvements in procedures and techniques which it knows are necessary, but for which neither time nor money can be found. It can puncture complacency and self-delusion.

10

QUALITY ASSURANCE AND THE CONTRACT

Contracts and system standards

Chapter 2 included a general description of the principles underlying the more common forms of contract encountered in the construction industry. It referred in particular to:

Architect's Conditions of Appointment
(Royal Institute of British Architects)
Conditions of Engagement
(Association of Consulting Engineers)
The JCT Standard Forms of Building Contract
(Joint Contracts Tribunal)
ICE Conditions of Contract
(Institution of Civil Engineers).

These forms of contract are produced by representative bodies and are generally accepted as establishing a reasonable balance between the conflicting interests of the various parties involved in a construction project. They are legal documents. They are couched in terms which can be interpreted with an acceptable degree of precision and many of their provisions have been tested in the courts over a number of years. Although many could argue that they are out-of-date and do not always function as well as one would like, they are a working system which everyone understands. They represent the *status quo*.

One of the functions of standard forms of contract is to provide purchasers with an adequate degree of confidence or assurance that the construction works designed or built on their behalf will satisfy their needs. It can therefore be argued that there is no need for purchasers to take any additional measures to protect their interests. Nevertheless, an increasing number of purchasers are requiring their suppliers to comply not only with conventional conditions of contract but also with the requirements of quality system standards. There is a variety of reasons why this action is being taken, and it includes:

1. The existence of legislation which requires that certain types of work should be carried out under a quality system complying with a specific standard. An example of this is the obligation on owners of nuclear installations to comply with standards such as BS 5882 or IAEA 50-C-QA as a condition of being granted a licence to operate.
2. The decision by a purchaser to operate a quality system of the type described in BS 5750 in respect of his own operations. In so doing, he would institute a procedure to review the quality systems of his suppliers (including architects, consulting engineers and contractors) and invoke compliance with an appropriate quality system standard in his procurement documents.
3. The conclusion by a purchaser that conventional contractual arrangements are inadequate to assure the quality of his purchase and that he will benefit from the additional powers which a quality system standard will confer upon him as purchaser and the additional obligations it will impose on designers and contractors as suppliers.

Such contractual requirements seem set to become increasingly commonplace. Indeed, one of the purposes of quality system standards is that they should be invoked in contracts. To quote from BS 5750:Part O, *Guide to the principal concepts and applications*, Section 1:

> 'In the contractual situation, the purchaser is interested in certain elements of the supplier's quality system which affect the supplier's ability to produce consistently the product or service to its requirements, and the associated risks. The purchaser therefore contractually requires that certain quality system elements be part of the supplier's quality system'.

The following are typical of clauses included in Invitations to Tender to specify the obligations of the contractor in respect of quality assurance:

1. The Contractor will develop and operate a formal Quality Assurance System in accordance with BS 5750:Part 2 and to the approval of the Supervising Officer (SO) and shall appoint and name a representative on-site to oversee its implementation.
2. The Quality Assurance System will cover management organization and procedures for the contract.

3. In the event of the Contractor being awarded the contract, he will submit within 4 weeks of receiving contract notification a complete version of his proposed Quality Plan to the SO for his approval.
4. The approved Quality Plan will be considered as forming part of the contract documentation and the Contractor will operate in accordance with its requirements. Any changes to the plan shall be agreed with the SO.

To examine the implications of such demands, let us consider some of the requirements of a typical quality system standard and compare them with the provisions of a typical form of contract. The standard selected is BS 5750:Part 2 *Quality systems — Specification for production and installation*. For comparison purposes, let us choose the JCT Standard Form of Building Contract for Local Authorities with Quantities.

SCOPE

BS 5750:Part 2 is designed to specify quality system requirements for use 'where a contract between two parties requires demonstration of a supplier's capability to control the processes that determine the acceptability of product supplied'.

The JCT form of contract does not require any specific demonstration of a supplier's capability. It merely stipulates that the Contractor is required to carry out and complete the works as defined in the contract documents using materials and workmanship as specified or to the reasonable satisfaction of the Architect or Supervising Officer (SO). The wise purchaser will very often take the elementary precaution of checking a contractor's capability before inviting him to tender, but any pre-qualification documents produced by a contractor in substantiation of a quality system are most unlikely to form part of the contract.

FIELD OF APPLICATION

BS 5750:Part 2 declares itself to be applicable when the following two conditions apply:

(a) the specified requirements for product are stated in terms of an established design or specification;
(b) confidence in product conformance can be attained by adequate demonstration of a certain supplier's capabilities in production and installation.

The first of these two conditions applies in conventional contractual arrangements. The specified requirements for the products of construction are usually voluminous, far more so than those of many other industries. We are all well used to this. The second condition, however, is an altogether different matter. Whereas in other industries there may be circumstances wherein compliance with specification can adequately be verified solely by the demonstration of a supplier's capability, to suppose that this can be the case on a construction site requires a quite unreasonable suspension of disbelief.

MANAGEMENT

Quality system standards require the establishment and definition of management structures for the assurance of quality. They stipulate, for example, that responsibilities for attaining quality objectives should be formally assigned; that there should be a management representative who, irrespective of other responsibilities, should have defined authority and responsibility for ensuring that the requirements of the standard are implemented; that systematic management reviews should be instituted, and so on.

In contrast, the JCT contract merely requires that the contractor 'shall constantly keep upon the works a competent person-in-charge'. The Architect or Supervising Officer is entitled (but not unreasonably or vexatiously) to 'issue instructions requiring the exclusion from the Works of any person employed thereon' but otherwise there is no constraint on the contractor's freedom to manage the site as he wishes.

PURCHASING

In comparison with the previous topic, BS 5750 says comparatively little about purchasing. It confirms the contractor's responsibility for ensuring that bought-in materials comply with specification and draws attention to the need for clarity in purchasing data. It also prescribes the right, if specified in the contract, of the purchaser or his representative to inspect at source or upon receipt.

The JCT form of contract is far more specific and detailed. It states that the contractor (or supplier) may not sub-contract work without permission and it endows the Architect/SO or his representative with substantial powers to sample and test materials and to reject those which are not satisfactory. It also requires that the Architect/SO or any person authorized by him shall have access to workshops or other places off the site from which materials or equipment are being obtained.

CONTROL OF WORK

In the language of quality system standards, construction work is a 'special process'. That is to say, it is a product which cannot be verified by inspection and testing in its final state and in which deficiencies may become apparent only after use by the customer. The means of control recommended for special processes in BS 5750:Part 2 is that they should be carried out under 'controlled conditions'. Examples given of 'controlled conditions' include documented work instructions and the certification of processes and equipment. It is also recommended that there should be 'monitoring'.

Unless supported by some other sanctions, these measures for the control of work are weak. The requirements of the JCT form of contract are far stronger and more explicit. There is provision for a Clerk of Works whose duties are to watch and supervise the works and to test and examine any materials used or workmanship employed. The Contractor may be compelled to uncover and open up works for examination and will have to bear the costs of such opening up and the subsequent making good if tests show that materials or workmanship do not comply with the contract. If he defaults by failing to proceed regularly and diligently, or by disobeying a legitimate Architect's written notice to remove defective work or improper materials, the Employer may determine the contract and engage another contractor to take over and complete the works. In this event the first contractor may be required to compensate the Employer for any loss or damage caused by the determination.

On the other hand, it can be argued that the obligations which quality system standards impose upon a supplier to prepare documented procedures for the execution of work, for inspection and testing and for the control of non-conformances, provide valuable additional weapons for the purchaser's armoury. This is particularly true if the procedures are subject to the approval of the purchaser and enforceable under the contract. The JCT conditions of contract do not provide such powers.

RECORDS

All quality system standards place emphasis on the need for documentation and the keeping of records. BS 5750 says that 'Quality records shall be maintained to demonstrate achievement of the required quality and the effective operation of the quality system'. Note the two-fold purpose of this requirement. Suppliers (contractors) are required to produce records to verify both the quality (i.e. compliance with specification) of the product, and the effectiveness of the system. By contrast, the JCT Conditions of Contract merely require that the contractor shall, 'upon request', provide 'vouchers to prove that the materials and goods' are 'so far as procurable of the respective kinds and standards described in the Contract Bills'. The principal verification of quality available to the purchaser derives from the supervision, inspections and tests carried out by or on behalf of the Architect/SO.

COMPARISON

The above examples expose a fundamental difference in philosophy between quality system standards and the JCT conditions of contract. Similar studies of the ICE conditions of contract used for civil engineering work lead to the same conclusion. Quality system standards are based on the premise that the quality of a product can adequately be assured by the appraisal of the management capabilities and techniques of the producer, followed by monitoring and the examination of documentary evidence of compliance. In contrast, construction contracts adopt a more sceptical, and perhaps a more old-fashioned approach. While responsibility for complying with specification is firmly placed with the supplier, the unspoken assumption is made that unless the purchaser maintains his own representation on the site to 'watch and supervise the works', the resultant structures or buildings will not be in conformance with requirements. The purchaser is therefore empowered to take an active role in controlling the quality of work, but is granted only minimal influence on the methods used or the management systems adopted. These are deemed to be the sole prerogative of the contractor since he carries both the financial risk and the responsibility for meeting the specification.

Which of these two philosophies is the more suited to construction work? The question is an important one since in many other industries the techniques of quality assurance have been startlingly successful in improving standards and reducing costs. Managements in these industries would not contemplate abandoning these techniques, and would probably regard the arrangements in the construction industry as archaic. Let us list some of the characteristics of classes of work which may be considered susceptible to control by system appraisal, monitoring and documentation:

1. Environmental conditions are controllable.
2. The work force is constant.
3. The techniques used do not change.
4. The materials used are of consistent quality.
5. The product consists of batches of identical articles capable of being sampled on a mathematically valid basis.
6. Work can be organized in such a way that all personnel will be motivated to do it properly.

Many of the tasks which form part of construction comply with these criteria. The production of reinforcing steel, the manufacture of trussed rafters, the laying of floor tiles and the testing of samples in a laboratory are examples which spring to mind. A purchaser, whether a client employing a main contractor, or a contractor employing a sub-contractor, could quite reasonably conclude that a system of audited self-certification or supervision by an approved third-party certification body would adequately protect his interests in respect of such work.

However, only a minority of the tasks undertaken on a typical construction site are capable of control solely by techniques of quality assurance. Tasks which are constantly changing, or which are weather dependent, or which rely on natural materials, or which are customarily carried out by casual labour, all require continuing supervision. This supervision can be supplied only by those who carry the financial risk if things go wrong. It is inconceivable at this stage in the industry's development that a contractor might say to himself: 'that bricklaying sub-contractor is certified by an approved body which has appraised his system and carries out random checks on his activities at six monthly intervals, therefore I will not myself attempt to supervise his work'.

So much for construction contracts. What of the contracts between clients and their professional advisers? These define the duties and obligations of architects and engineers and establish the terms of payment. They do not address the possibility of failure to perform, neither do they require that designs should have fitness for purpose. All that is required of the professional is that he should exercise 'reasonable skill and care' in his work. There are no contractual sanctions available to the purchaser in the event of default other than termination of the engagement after a reasonable period of notice.

This is not an unreasonable arrangement. The most valued asset of a consultant is his professional reputation, and the power to threaten loss of reputation is a strong sanction in the hands of the purchaser. Moreover, consultants have not traditionally been required to compete on price (although this practice is changing). Unlike contractors, they have not so often been tempted to compromise their integrity for the sake of short-term profit. It could well be argued that they offer a quality of work which is adequately assured by their reputation and that further measures such as purchaser supervision or audit are quite unnecessary. On the other hand, the high proportion of construction defects which are found to have their origin in the design office is perhaps an indication that not all is well. Furthermore, the growth in fee competition for consultancy work will inevitably increase the pressure on designers to reduce the costs of design even if this leads to an increase in total costs, either through errors or because designs which are cheap to produce may be more expensive to build and maintain.

The contractual options

One is forced to conclude that while quality system standards and typical conditions of contract may share many of the same objectives, they follow quite different philosophies in the methods chosen to achieve these objectives. Both philosophies have their merits. How can clients select the best of both worlds to improve standards of construction performance? Should they, for example, include a requirement for compliance with a quality system specification in their purchase orders or contracts?

As mentioned earlier in this chapter, section 0.1 of BS 5750 would appear to advise that they should. Section 0.2, however, leaves the option open by suggesting a number of ways in which a supplier may provide assurance of compliance with specification and concludes:

> 'The assurance provisions should be commensurate with the needs of the purchaser's business and should avoid unnecessary costs. In certain cases, formal quality assurance systems may be involved'.

Suppose a purchaser does make compliance with a system standard a condition of contract, how does he enforce this requirement? One option may be to use only those suppliers who are registered under accredited third-party certification schemes. For such suppliers, enforcement of compliance is undertaken by the certification bodies and there need be very little change in the contractual relationship between purchaser and supplier. All the purchaser needs to do is make continued registration by the certification body a contractual requirement. He may also require suppliers to put up some form of bond which would become forfeit in the event of de-registration. It could be argued that, under such an arrangement, any form of supervision by or on behalf of the purchaser would become superfluous since compliance with specification would be assured by the surveillance activities of the certification body. On the other hand, it needs to be borne in mind that certification against a system standard establishes a supplier's or contractor's capability. But capability alone will not guarantee performance, particularly in the construction industry. A certification body would not be able to maintain a permanent presence on the site, nor could it resolve design problems, nor would it be in a position to certify work for payment. It therefore seems most unlikely that the role on site of the Engineer or the Architect/Supervising Officer will disappear, although it is conceivable that there could be a reduction in the numbers of inspectors or clerks of works deployed on the purchaser's behalf.

In the event that there are no suppliers of a given product or service who are registered under a suitable accredited third-party certification scheme, a purchaser may decide simply to include in his tender documents a clause requiring compliance with a standard such as BS 5750. Such action can have its pitfalls. While BS 5750 is quite specific in the obligations it imposes on suppliers, a purchaser's ability to enforce these requirements will depend on the powers vested in him or his representative through the terms of the contract. The differences in philosophy and detail between standards and conventional conditions of contract which were discussed earlier in this chapter, and the need for interpretation of the standard's requirements in the context of the particular class of work, can make this a difficult and tortuous arrangement and it is not to be recommended.

To provide a better means of enforcement, some purchasers adopt the approach exemplified on p. 160 and require that all the contractor's or designer's procedures and documentation should be to their approval. They may go further and give themselves the right to audit supplier's systems and to impose corrective actions if these audits reveal departures from the standard. A supplier who fails to follow approved procedures, or who ignores demands for corrective action may then find himself in breach of contract and subject to all the sanctions contained in the contract for this purpose.

Such an approach represents a significant change in contractual philosophy and could have far-reaching effects. The purchaser is taking power, not just to satisfy himself that the works he is buying comply with his requirements, but also to control the organization structures, the management systems and the working procedures of all those who contribute to their design and construction. There is likely to be conflict between the quality system operated by the supplier to meet his internal objectives and that imposed by the purchaser. Quality system standards are open to more than one interpretation and they include requirements which, while justifiable in particular cases, could if applied out of context, substantially increase a contractor's or designer's costs without compensating benefits. Such a move would be detrimental to the interests of contractors who could not be expected to price works over which they would not have control and is likely to be equally unwelcome to professional architects and engineers. It could also have profound implications on the apportionment of risk and of liability.

So, what actions should the discerning client take to improve confidence that the construction works he is about to purchase will satisfy his requirements? The following guidance is offered:

1. Require evidence of an effective quality system as a condition of pre-qualification. This applies to design consultants as well as construction contractors, particularly in circumstances of fee competition. In the event that there are too few tenderers with adequate quality systems to form a

market, be prepared to load the price of unqualified tenderers to compensate for the additional work which will have to be done by someone else to provide an equivalent level of assurance. Do not assume that a contractor will be able to set up a quality system on your project on a 'one-off' basis. If he believes in quality assurance, he will have a company-wide system already. If he does not believe in it, no system he sets up will be effective.

2. Take advantage of existing accredited third-party certification schemes for contractors and suppliers, and specify their use whenever possible.
3. Do not make compliance with a quality system standard a condition of contract.
4. If additional assurance beyond that provided by the contract is required, select clauses of a quality system standard that are appropriate, and include them as 'special conditions of contract'. For example, there may be elements of work for which a contractor should produce formal procedures subject to the approval of the Architect or Engineer; there may be special requirements relating to records and documentation; the purchaser may wish to institute a system of quality audits; there may be particular requirements for the control of sub-contractors and suppliers, and so on. If this is the case, spell it out so that the contractor can allow for any additional costs in his tender.
5. Be wary of imposing conditions of contract which restrict the contractor's obligation and right to manage the site and to select the most economic methods and techniques for achieving compliance with specification. He carries the commercial risk and cannot be expected to give up the reins of control.
6. In conclusion, do not expect quality assurance measures to be effective unless you, as client, have staff experienced in their application in the context of construction. Quality assurance will save you money if it is applied properly. It will cost you money if it is not.

11

A STRATEGY FOR SUCCESS

We have now examined all the main elements of a quality management system for construction work. The final task is to put it all together and make it work. The Bible tells the parable of the foolish virgins, who got it wrong, and the wise virgins, who got it right. A similar parable can illustrate how to get things wrong, or alternatively right, on a construction project. Let us consider two scenarios.

The foolish way

THE START

Picture day one of a two-year programme for a construction project. The contract has been awarded after competitive tendering on the basis of drawings and specifications produced by a reputable architect assisted by an equally well-known structural engineer. The client is a fast-growing commercial organization who has made it very clear to the architect that he wants his project completed on schedule and within his budget.

The contractor's construction director was relieved to have been awarded the contract. His company had been going through a lean period and, but for this success, would have had to institute a programme of staff redundancies. There were, however, some small clouds on the horizon. One was the fact that his price of £12 million was about £500,000 below that of the next lowest bidder. On going through the tender make-up again, he had discovered some errors and misjudgments which, when costed out, showed that at the price awarded there would be no profit left in the job. All those involved in the tender process had been suitably admonished. They seemed to take it philosophically — it was probably a change from being told that coming second was not good enough and that if they did not 'sharpen their pencils and get some work' they could expect to have to look for employment elsewhere.

However, price was not the only problem troubling the construction director. This was a larger project than any his company had tackled before and contained some technically complex work of a type unfamiliar to his staff. He knew that he had only been included on the tender list because of some rather questionable arrangements which he would prefer not to go into too deeply. Still, with a bit of luck he would be able to get some good sub-contract prices for the awkward bits. The fact that he had allowed very little in his price for technical supervision was perhaps unfortunate, but the building inspectors and clerks of works could probably be relied upon to prevent anything terrible happening.

The Architect too, was worried. The client had been difficult during the pre-tender stage. He kept changing his mind and had refused to co-operate in the preparation of a documented design brief. When the Architect had attempted to confirm his understanding of what was required, the client either changed his mind again or failed to respond altogether. The Architect was also worried about the contractor. He was sure that he did not really understand the technicalities of the work and that this was the reason for his low price. The Architect had in fact recommended that his bid be disqualified, but the client did not see why he should pay £0.5 million more than was necessary, and he had had no support from his senior partner when he discussed the matter with him. It was a mystery how the contractor got on the tender list in the first place.

Then there was the matter of the site investigation. The structure required the construction of a deep basement. The structural engineer had had some boreholes sunk to check the ground conditions, but when the report came in there were one or two points which caused him concern. It appeared there was a possibility of running sand in the north-west corner. The structural engineer had wanted more boreholes, but the client had refused to sanction any more money. The site investigation budget had been used up and, in any case, there was no time. Any problems would have to be sorted out during construction.

The contractor's Agent had mixed feelings. He was new to the company and he was determined to make a name for himself. He was aware that his performance would be watched carefully. His company had recently implemented a new computerized system for monitoring progress and any failure on his part to achieve cost or programme targets would bring immediate and unpleasant retribution from Head Office. He was aware, too, of the technical problems ahead of him. The basement excavation might prove tricky, but his quantity surveyor reckoned that it should be worth a few hundred thousand pounds in claims. In this and other areas, the Architect had obviously had to cut corners to meet the client's budget. Some of the drainage arrangements and waterproofing details to the multi-storey car park above the supermarket seemed quite inadequate. There might be some claims there too. In any case, he was well aware that his superiors regarded technical details to be of little interest. The important thing was to make a profit, and although the board paid regular lip service to the concepts of 'quality' and 'customer satisfaction', he did not take such exhortations seriously.

HALF-WAY

After a year, the project was already two months behind programme. Most of the delay occurred during construction of the basement. The structural engineer's suspicion of running sands in the north-west corner had proved only too true. The contractor eventually overcame the problem, but claimed that his costs in so doing should be borne by the client as the scale of the problem was not made apparent by the soils investigation and could not reasonably have been foreseen. Far from accepting any responsibility himself, the client appeared to hold the Architect to be at fault. This was not the only cross the Architect had to bear. The contractor had written to him implying that some of his drainage and waterproofing details were inadequate. Given the standard of work produced by some of the sub-contractors, this was a bit rich, and probably just a ploy to cause him embarrassment. He had replied in no uncertain fashion that the contractor's job was to build the works in accordance with the drawings, and that if he wanted to put in any more damp-proof courses or additional drainage, it would be at his own expense.

Although the contractor hoped to be reimbursed for his additional work on the basement, the debts were hurting his cash flow. The underpricing was also beginning to show, and the sub-contractors who had been prepared to take the work at prices he could afford were not performing well. One had already gone bankrupt, and another was threatening to do so if obliged to perform according to the contract specifications. The Agent decided that it would be best not to press them too hard. In any case he had no staff to supervise them. He himself was far too busy chasing the Architect for information and working on the claims to spend time walking around the site. There were times when it was better not to know.

The contractor's new financial control system appeared to be proving its worth. It had signalled potential losses in the first few months, but suitable pressure had been applied to the Agent and he was now producing more satisfactory figures. Clearly he had learnt where his employer's priorities lay. All in all, the project was not working out too badly, certainly the losses would not be as serious as the construction director had earlier feared. It had been decided not to make an issue of the waterproofing details since the company was hoping that the Architect would recommend them for another contract for which he was responsible, and it would be a mistake to do anything that might lose his favour.

TEN YEARS LATER

The project reached substantial completion about six months late. Smarting at the delay and consequent loss of rental, the client took possession and his tenants started to move in. The contractor's books showed a small profit but this was whittled down almost to nothing during the Defects Liability period by the need to maintain a team of tradesmen on the site to carry out remedial works, most of which were due to poor workmanship by sub-contractors who had long since left the scene. However, eventually the work was finished, the claims were settled and the Architect issued the Final Certificate. The eventual Contract Sum was £1.5 million above the tendered figure which angered the Employer, but he had no option but to pay up.

After a few months, rain-water started to leak into the supermarket below the multi-storey car park. The car park itself was also showing signs of trouble. It was clad in brick because it was in a prestige location and the client wanted a 'quality' image. Unfortunately the brickwork was showing signs of stress and was becoming badly stained where rain-water discharging from the top deck cascaded down the outer walls. Some of the courses of brickwork were breaking loose and the local vandals were lobbing bricks into the car dealer's yard next door. They then added graffiti to complete the air of neglect. The contractor was called back, and with bad grace he plugged the leaks, cleaned off the encrustations of lime and repaired the brickwork. But more than mere patching was needed.

Three years after assuming occupation, the tenant of the supermarket, his patience now exhausted, made a formal complaint to the building owner about the ingress of rain-water into his store and threatened legal action. The owner, stung at last to action, engaged a consulting engineer to examine the building and advise on what action should be taken. The consultant's report was devastating. He found that the waterproofing membrane to the top floor of the multi-storey car park was defective, that the arrangements for shedding rain-water were quite inadequate and that there were insufficient movement joints to provide for the expansion and contraction of the expensive brickwork cladding. In addition to these design faults, his investigations showed widespread bad workmanship. Many of the damp-proof courses were badly laid. Brick-ties were missing or incorrectly fixed. In some of the few brickwork movement joints which had been provided, plywood sheets used for forming the joints had been left in place, accentuating still further the already inadequate arrangements for coping with expansion and contraction. The total cost of rectification was estimated to be £500,000.

There followed a further period of minor repairs and patching, causing continued inconvenience and discomfort to the users without curing any of the underlying defects. Finally, five years after taking possession, the client issued a writ for compensation, naming the contractor, the architect and the structural engineer as joint defendants. In the protracted arguments which ensued, the architect claimed that it was not his job to design every single detail in the building, nor could he be expected to supervise the contractor's men all the time. He had used 'reasonable skill and care' and this was all he had contracted to do. The contractor admitted faults in workmanship but pleaded that in the main he had built what was on the drawings after having pointed out to the architect that there were details that did not comply with accepted practice. The structural engineer said that the cladding and drainage had nothing to do with him; he had only designed the structural framework.

Eventually an out-of-court settlement was agreed whereby the plaintiff would be awarded the sum of £300,000, of which the architect and contractor would pay 40% each and the structural engineer would pay 20%. Had the settlement not been reached and had the case gone to court, the legal costs alone could have reached much the same amount. The owner had hoped for more, but he realized that the quicker a settlement was made, the better. He was right — a year later he might have received nothing, since by then the structural engineer was dead, the architect had gone abroad and the contractor had filed for bankruptcy.

Eight years after completion, the building was again shrouded in scaffolding as another contractor undertook the major repairs which were necessary to correct the bad design and bad workmanship which had plagued it from the start. What had been intended as a prestige city-centre development bringing in high rents to the owners and giving comfort and convenience to the public had become instead an ugly and dilapidated slum, rejected and vandalized by those whom it was intended to serve.

The better way

THE START

The client had spent many years and a great deal of money in developing his plans, in assembling the project finance and in purchasing the site. He was under considerable pressure to start construction as quickly as possible. However, his past experience led him to resist the temptation to rush into things too fast. He knew the catastrophic effects which late design changes can have on a construction programme. His first act was to appoint a senior member of his own staff to act as the Client's Representative. The person selected had a thorough knowledge of the requirements of the new project. He had the full confidence of other senior management within the company and was given the authority to act for the Client in resolving the day-to-day problems which would arise during construction. He was told to work closely with the Architect to make sure that a precise and comprehensive design brief was developed, and to be the main point of contact with the construction contractor soon to be appointed.

The Client's Representative's first task was to select and engage the Architect. The firm chosen had a good local reputation. They were also able to demonstrate that they operated an effective quality management system. On their recommendation, a firm of structural engineers was appointed to design the foundation and structural frame. The structural engineer was quick to specify and order a site investigation. This proved more expensive than expected because some of the strata exposed required further examination. The Client's Representative was prepared to sanction the additional expenditure rather than risk claims for unforeseen ground conditions at a later stage.

The Client's Representative advised the Architect that tenders for construction should be invited only from a prequalified list of contractors and that the prequalification process should include checks on financial status, track record, staff resources and quality systems. The lowest valid tender of £12.5 million was just within budget and was therefore accepted. It was a requirement of the contract that the contractor should institute a series of internal quality audits of his work, the first to be carried out one month after entry to the site and the remainder at six month intervals thereafter. The prequalification enquiries had shown that the selected contractor had suitable auditing procedures and staff qualified to put them into effect. Reports of audits, including details of resultant corrective actions, had to be copied to the Client's Representative.

As part of the tender planning, the Agent-designate, together with the company chief engineer, had prepared a first draft of the project quality plan. The purpose of this was to establish staffing requirements for quality management and to identify particular elements of the work which could be expected to lead to quality problems and for which written work instructions would be required. Some of these elements required further study so that outline solutions could be arrived at for pricing purposes. In cases where the work element concerned would be sub-contracted, it was necessary to advise potential sub-contractors that they would be required to prepare written work instructions for the contractor's approval and that they should make allowance for so doing in their tenders.

After the work was awarded, the quality plan was brought up to date, responsibilities for quality management were defined and allocated, and arrangements were made for the work instructions to be finalized and issued. The plan also identified the inspection and testing regimes which would be implemented (including tests to be carried out by sub-contractors), the record documents to be produced and the programme of audits to be held in order to verify the effective functioning of the system. Finally, in recognition of the fact that the design work was still incomplete, arrangements were made for updating and amending the quality plan on a regular basis.

HALF-WAY

The first issue of the contractor's quality plan was made just before he moved on to the site. One of the work instructions scheduled in the plan related to the installation of brick cladding. While preparing this instruction, the engineer to whom the work had been delegated became concerned at the lack of movement joints. He brought the matter to the attention of the company chief engineer who had wide experience of the problems associated with brick clad buildings. He concurred that the joints were inadequate, and wrote a courteous but well-argued letter to the Architect suggesting that he should re-examine his proposals. After discussing the matter with his structural engineer, the Architect accepted the contractor's arguments and made the necessary changes to his drawings. Being a much smaller organization, the Architect's office operated a less formal quality system than that of the contractor. Nevertheless, the senior partner made it a rule to carry out his own regular reviews of work done by his more junior partners. In the course of his first review, he examined the arrangements for waterproofing and draining the top deck. He came to the conclusion that the falls were not steep enough to prevent ponding, that the sealing of construction and movement joints was inadequate and that the drains and down pipes were liable to become clogged with rubbish. Fortunately these matters were discovered well in advance of the work being carried out, so the necessary changes to the drawings were made at minimal expense.

The contractor's Agent, who had only recently joined the company, was sceptical about the quality system. Some aspects he was prepared to accept: the quality plan, for example, seemed to be a useful discipline for preventing defective work. He was, however, far from happy at the thought of his project being audited. This seemed to him a quite intolerable invasion of his territorial rights. Before commencing the first audit, the auditors took pains to reassure him that the purpose of the audit was to evaluate the effectiveness of the company's systems, not its people. It was not intended to point a finger of blame at him or at any of his staff. If the systems were inadequate or were not being implemented, it was to everybody's advantage that the facts should be known so that they could be modified if necessary. The auditors had no power to require him, or anybody else, to do their bidding. All they would do would be to draw attention to what seemed to them to be evidence of system breakdown. It would be up to the management to decide whether action was needed and what that action would be.

The first audit duly took place on schedule one month after the granting of access to the site. Since the project was still at a very early stage the audit took only one day and the Audit Report was brief. It established that an acceptable quality plan had been issued and that responsibilities for technical supervision had been allocated. It drew attention to the fact that the company's own planning procedures were not being followed — the planning engineer was new to the company and had not been issued with a copy of the planning manual so he was using the system he had become used to with his former employer. It was also found that the technical documentation attached to some of the sub-contract enquiries failed to give sufficiently clear instructions on the quality control activities to be performed by the sub-contractors. A copy of the Audit Report together with details of the corrective actions proposed by site management was sent to the Client's Representative.

The next audit took place six months later. By then the project had settled down. The quality plan had been re-issued twice in the light of new design information emanating from the Architect's Office. All the work instructions had been issued and were being followed. The auditors queried one point however: they observed that payments to sub-contractors were being certified by the quantity surveyor on the basis of a quantitative measure only and that this had led to instances of payment for non-conforming work. The site was operating a new procedure whereby critical items of work were formally checked and signed off by the project engineer and the clerk of works. It was suggested that copies of the resultant documentation should be sent to the quantity surveyor. He would then be able to verify that work was satisfactory before he certified payment, and many of the measurements he had formerly undertaken would become unnecessary.

By the time the half-way audit was due, the Agent was becoming reconciled to the concept. The findings of the first two audits had not been used as weapons of retribution, indeed in the discussions which took place afterwards, he was able to have his say on a number of the system improvements which were considered. Nobody had been humiliated, the auditors had been courteous and professional in their approach to him and to his staff, and his position as Agent was respected. Quality auditing had become as much a routine part of life as financial auditing.

TEN YEARS LATER

The project was completed on time, the client took possession and his tenants moved in. A few small repairs were made during the Defects Liability period. Approved contractor's claims for extra work totalling £0.5 million raised the Contract Sum to £13 million.

The building functioned as it should. There were no leaks, no litigation and no bankruptcies. The legal profession was denied its pound of flesh.

The choice

What conclusions are to be drawn from these two hypothetical case studies? Firstly, the quality problems of the construction industry seldom arise from new or intricate technology. Most of the defects recorded in 'the foolish way' were simple matters, well within the state of the art. Secondly the common causes of poor quality are greed, idleness and ignorance. In 'the foolish way' the client failed to brief his architect properly and thought that the way to have his project built at minimum cost was to let the work to the lowest bidder. The architect was weak in his approach to his client and vain in dealing with the contractor. The contractor was desperate for work, badly organized and lacking in scruple. Yet each of these parties, at the time, believed that what they were doing made good commercial sense.

There is a variation of Murphy's Law which states 'Once you have made a mess of things, anything you may do to put them right will only make them worse'. Trying to put things right after the event is not only unlikely to be effective, it is also costly. Table 11.1 puts some figures to the costs of the two

Table 11.1 The cost of getting it right

Item	The foolish way	The better way
	£m	£m
Tender sum	12.0	12.5
Additional site investigation	—	0.1
Unanticipated extras	1.5	0.5
Loss of rental	0.3	—
Remedial works	0.5	—
Legal costs	0.1	—
Sub-total	14.4	13.1
Less compensation received	0.3	—
Total project cost	14.1	13.1

scenarios we have considered. So, the first Client ended up £1m worse off than the second in direct costs alone. He also suffered substantial indirect costs in respect of wasted management time, loss of goodwill and disruption. Likewise, although the first contractor managed to negotiate £1m more in claims than the second, his profits were quickly swallowed up in defect rectification, legal costs and compensation. The second contractor not only made more money, he came out of the contract with an improved reputation, a satisfied client and the prospect of negotiated repeat orders. Similar comparisons can be made between the first and second architects.

These examples are, of course, deliberately contrived to make the point that an organization can serve its own self-interest by establishing effective systems of quality management and encouraging its suppliers to do likewise. If the participants in the second scenario are presented as unreasonably more intelligent and more virtuous than those in the first, the author hopes that this will be accepted as a pardonable exaggeration, but one which can be justified by the fact that quality systems do, in fact, encourage employees to adopt an intelligent and conscientious approach to their work. They can initiate a 'virtuous circle' whereby attitudes, systems and skills combine to develop a momentum of improvement to the benefit of everyone.

Let us, however, leave our hypothetical scenarios and return to the hard and dirty world of construction. It has to be faced that the concepts and theories of quality management, with their emphasis on planning, control and verification, do not have an immediate appeal to many of those who spend their lives building large structures. There is a machismo which attaches to the virile construction boss who can sense what is wrong with a project and then wrest it back from the brink of disaster. This, to many, is what makes life worth living, and it is far more exciting than planning, or developing procedures, or assembling records.

Any construction organization contemplating the establishment of a quality system, whether it be to achieve certification by an accredited body, to enable it to pre-qualify for particular types of work or simply to improve its own efficiency, faces a long and difficult path. The fact that the theories of quality management can lead to successful and profitable operations in factories cuts little ice in the minds of those steeped in the traditions of construction. They feel in their bones that they differ from and, dare one say it, are superior to people who work in factories. They are probably right, but if it is true that the management of construction sites is a uniquely difficult task, it is all the more important that those who are charged with the task should have access to the latest and most effective management techniques.

These are all matters which warrant the careful consideration of top management. There is, however, a final and fundamental question which needs to be addressed: Does the organization want to run its business on a basis of integrity and compliance with the rules, or does it wish to pursue short-term profits by tolerating bad practices and short-changing its clients? If the answer is the latter, then any quality system which might be established will be a sham, the people operating it will be frustrated and it will be an additional cost burden. Furthermore it will not deceive a client or a certification body. On the other hand, if it is management policy to get things right and to require its employees to adopt an honest and conscientious attitude to their work, a formal quality system will provide an excellent and economic means of so doing. What are the steps which must be taken to achieve this objective?

The company ethic

Management gets the employees which it deserves. If there is laxity or dishonesty at the top of an organization, or if people perceive this to be the case, sooner or later there will be laxity or dishonesty at other levels as well. This will inevitably lead to poor quality and customer dissatisfaction. Conversely, staff respect honesty and integrity in their managers and will tend to apply similar standards in the conduct of their own affairs. To set the standard, the chief executive must make a public statement of his personal attitude in these matters.

In 1978, the president of the Japanese construction company Kajima Corporation announced in his inauguration address his intention to introduce a company wide system of 'TQC' (Total Quality Control). The purpose behind this innovation was defined as follows: *

1. Enhancement of the spiritual resolve of each individual to serve the needs of society through the company's vigorous entrepreneurial activities.
2. Elevation of the spirit of loyalty to the company's as well as society's needs and prosperity.
3. To build up a healthy and capable entrepreneurial entity to ensure the long lasting prosperity of the company as well as the society under which the company operates.
4. To get the whole organization united as a body to realize the foregoing basic policy.

Such sentiments may appear a trifle pious to those used to British understatement, but they are typical of the policy declarations of Japanese companies who have become world leaders largely through meticulous

* Extract from paper 'Total Quality Control and Quality Assurance in the Construction Industry' presented in Singapore, April 1985.

attention to the quality of their products. On issues such as these, there is no harm in management wearing its heart on its sleeve, but it should beware of humbug and not pretend to policies it has no intention of implementing.

The scholar Abraham Maslow once examined the effects of various stimuli and their powers to satisfy and motivate human beings. He concluded that the first and most basic needs are for food and shelter to ensure survival. Once these have been satisfied, people seek escape from danger or anxiety. Having achieved security they pursue the friendship and esteem of others. Finally there is the ultimate satisfaction of personal accomplishment and self-realization. A similar 'hierarchy of needs' can be discerned in organizations. If survival is at stake, profit is the most powerful motivator. Companies which are not faced with impending bankruptcy or extinction, and thankfully this is the majority, do not need to pursue the profit motive to the exclusion of all others. Yet, if asked to define the fundamental purpose of their business, the immediate response of many boards of directors would be that it is to maximize profit. If this were true, they would not be in the construction business. There are many easier and more secure methods of making money than competing for construction contracts. Some of them are legal.

Even if the pursuit of profit were to be an organization's most important goal, there is the question of time scale. Is the aim for profit to be short, medium or long term? The ethics of a labour-only sub-contractor faced with the overriding imperative of paying the wages at the end of the week can be expected to differ significantly from those of a mature business, dependent upon the continued loyalty of its work force and with a reputation to maintain. In its Quality Manual (Appendix B), Alias Construction quotes the corporate objective of its parent group (para 2.1). This recognizes the basic needs for survival and security ('to maintain and improve upon its position as a leading construction contractor and builder of houses') but then sets out three conditions which have to be met to satisfy these needs: reputation, the welfare of the work-force and profitability. These are goals to which the work-force can respond. If profitability is management's only objective, it can not complain if the work force follows its lead and works only to maximize its own income. This can be a short cut to disaster.

These are matters of considerable interest to potential purchasers who need to know the kind of organization they are dealing with. Presentation of the corporate ethic is an important part of marketing strategy. Companies adopt many ruses to demonstrate their position in the 'hierarchy of needs'. They contribute to charities, they sponsor the arts or sporting events, they endow university chairs, and so on. These activities win esteem and instil a feeling of personal pride in their employees. They provide evidence that the company has developed beyond the survival stage and can be relied upon to go about its business in a mature and responsible fashion.

The management commitment

So, the first step in the establishment of a quality system is to establish the philosophy or set of beliefs which will provide a sound foundation for what is to come. This requires action by the Chief Executive and his board of directors since these are matters which only they can articulate. But their involvement does not end here. Having nailed their flag to the mast, they have to provide sustained leadership or their fine words will be wasted. Quality systems can be implemented only from the top down. The process requires that long-established customs be examined and perhaps abandoned. Skeletons have to be brought out of cupboards and vested interests have to be challenged. Inevitably many people will find this process disturbing and uncomfortable. When the going gets rough, as it will, management must maintain the momentum.

The magnitude of the task must influence the choice of people appointed to tackle it. Many organizations make the mistake of delegating the leading role to someone approaching retirement or to a candidate for sideways promotion. Others see it as an appointment to be reserved only for those with a technical background. It is an error to assume that an expert in quality control can easily make the transition to quality management. Indeed, the reverse is often true, and such appointments carry a further disadvantage in that they can reinforce the misapprehension in the minds of other managers that quality management is a function of technical specialists. This is not so; there is no reason why people trained in accountancy, law or even quantity surveying should not become effective quality managers.

The selection of the quality system manager will inevitably be viewed by the work-force as a measure of management's interest and commitment to the project. The task requires energy, good communication skills and a knowledge of the business. People with these characteristics are in short supply and those who have them may well be reluctant to enter what may appear to be a career cul-de-sac. To overcome such difficulties, it is suggested that the task should be presented as a five-year project in management development. Not many companies will be able to establish an effective system in less than this time span and by limiting the length of the appointment the commission becomes more attractive to those possessing the necessary dynamism and ambition.

It is wise at this stage to establish the priorities to be followed in setting up the system. Is the main purpose to enable the organization to demonstrate compliance with a particular system standard, such as BS 5750? If so, will the organization seek third-party certification? Are these, on the other hand, more distant objectives to be considered only after an effective and economic system has been established? The answer to these questions can best be determined by studying the needs of the purchaser. If the market requires compliance with a system standard and accepts third party certification as adequate proof of compliance, then certification should be the main objective. Conversely, if purchasers prefer to select their suppliers on the basis of open competitive tender and to rely upon traditional contractual arrangements for the assurance of quality, then the requirements of system standards must take second place to the commercial needs of the company.

The appointed system manager will require a written brief. This should set out the purpose of the appointment and establish the ground rules to be observed by the manager himself and those upon whose co-operation he will have to rely. In preparing the brief, it needs to be borne in mind that the system manager should not be put in a position where he might usurp, or appear to usurp, the powers and obligations of line management. While it will be his responsibility to make recommendations that certain actions be taken, the authority to determine how and by whom the recommendations should be put into effect, and to give instructions, must remain with the responsible line managers. However, before he can make recommendations, the system manager will need facts and to obtain these he will have to ask questions. The power to call for information, and to probe the truth of this information, is the most important weapon in a quality system manager's armoury.

The conversion

Let us assume that the first steps have been taken. The corporate policy has been formulated and made public; senior management has decreed that a quality system to give effect to the policy will be established, and a management team has been appointed to make it all happen. What comes next?

Even the most well-managed organizations have difficulty in bringing about internal changes. This is particularly true of the construction industry where traditional attitudes tend to be deeply entrenched. The announcement of policy and the declaration of intent are only a start — they have to be followed by a sustained programme of explanation, education and persuasion in order to win the willing co-operation of those who will be affected. The quality manager has to be prepared to argue his case on its merits. If he cannot do this, then the possession of a mandate from the top will be of limited use to him. The corporate body will treat the quality system as a transplant from an incompatible donor and will reject it.

The strategy of the campaign of persuasion will be influenced by the organization's objective in establishing the system. If the primary purpose is to qualify for registration by a certification body, the role of the system manager will be to describe and explain to line managers the measures which the certification body requires and to assist in their implementation. The motivation for reform will arise from the decision to apply for certification, and provided the chief executive proclaims his intention with force and clarity, this should suffice to overcome any opposition.

The development of a quality system to improve the internal efficiency of an organization is a more difficult task. When opposition to reform is voiced, it will not be possible for the system manager merely to respond: 'That is what the certification body requires us to do. It may not make sense, and it may even cost us money, but that is not the point. We have decided to go for certification and if this reform is a condition of achieving this objective, it has to be done'. In the absence of such an incentive the quality system manager will have to be prepared to argue his case in respect of every move he makes. This can be an arduous and time-consuming activity but, in the long term, it can result in a better system, since every change which is made will arise from an analysis of the real needs of the organization, and not from a desire to satisfy the whims of an outside body.

Here are some of the arguments which a quality system manager can expect to encounter:

> 'Our quality performance is no worse than that of our competitors. The measures you suggest are therefore unnecessary'.

This statement discloses a state of complacency which it is necessary to puncture. The quality manager can pre-empt such arguments by commencing his presentation with a short video or tape-slide sequence containing examples of poor quality work culled from the recent history of the organization. If he can back these up with the associated costs of rectification, this will make the message even more telling.

> 'Even if there is scope for improvement in our quality performance, the system you propose will not be effective.'

This is a more difficult argument to counter. One can cite cases where quality systems have been successful elsewhere, but people tend to see their own circumstances as different from those of anyone else and therefore do not accept such comparisons. The best approach is to discuss each of the elements of a quality system in the context of the work carried out by the organization in question. If a particular element has no relevance, this should be acknowledged. The likelihood is that most of the requirements can be presented as examples of normal good management practice which, in the heat of battle, tend unfortunately to be neglected. This neglect gives rise to problems which then escalate the heat of battle, leaving even less time for good housekeeping and the prevention of defects.

Another device is to suggest that in the case of the examples of defective work shown previously, the senior management responsible could not have been aware of what was being perpetrated on their behalf, otherwise they would have taken action to stop it. A quality system would at least have provided a mechanism which would have ensured that they were informed.

> 'What counts in our industry is people. If the people are good, you do not need systems.'

This proposition is superficially attractive, but does not withstand close examination. The first sentence is clearly correct: a well-trained and enthusiastic team is essential if any worthwhile objective is to be achieved. However, people who are so gifted that they can operate without any system, or who can create their own systems as they go along, are extremely rare. To have a whole team with such talent would be wasteful and probably end in conflict.

Most human beings are not especially gifted. They forget things and make mistakes. They sometimes arrive late or fall ill. These shortcomings can be minimized if they work in an orderly environment and it is the task of management to create this environment through the establishment of systems appropriate to the work in hand. Managements who neglect this obligation are not only failing in their duties to their employers and their work-forces, they are encouraging the proliferation of people who are able to claim indispensability because no one knows what they do or how they do it. These are often the very persons who raise this particular argument.

The establishment of a quality system need not stifle initiative or require blind adherence to rules which have no point. It can enhance the performance of those of only average talent and enable genius to shine even more brightly. Even Mozart had to follow the rules of harmony and composition.

> 'What you are suggesting will add to our overheads and make us uncompetitive. We cannot afford it.'

This is an argument so fundamental that it needs a section to itself.

The costs

Very few organizations measure the costs of poor quality. It is therefore exceedingly difficult to prove that systems for the prevention of defective work are cost-effective. Whereas the direct costs of a quality system can be quantified with some precision (the salaries of the system manager and his staff, the costs of documentation, expenditure on audits, and so on), the corresponding benefits are far more difficult to assess. However, to counter the argument that quality systems cost more than they save, some assessment has to be made. Here are some of the items which contribute to the costs of poor quality:

Repair of defective work.
Purchase of replacement materials and components.
Delay or disruption while repairs are carried out or replacements obtained.
Handling of customer complaints.
After-sales remedial works.
Wastage of marketing effort.
Legal representation.
Court costs.
Compensation payments.

Companies which have seriously attempted to quantify these costs usually arrive at totals varying between 10% and 40% of turnover. By comparison, the typical cost of a quality system for conventional construction work is likely to lie in fractions of 1%. Even in extreme cases, such as off-shore or nuclear construction, it is unlikely to exceed 5%. So, even if the quality system can prevent only a small proportion of the costs of non-quality, the potential for savings is enormous. But, how can this be proved?

The reason why so few organizations are aware of the value of resources wasted as a result of poor quality is the human instinct of self-protection. For fear of retribution, the costs of having to do things twice, or of re-ordering wasted materials, are spread elsewhere. Sometimes materials are deliberately over-ordered in anticipation of waste: if the waste does not happen they are thrown away or pilfered. The effect of these actions is to conceal from senior management the long-term chronic problems within the organization which are sapping its strength. Instead of directing attention to areas where genuine savings can be made, the cost figures will just show that labour or material costs are rising or that overheads are too high. To respond to these indicators merely by, for example, withholding a wage increase or seeking cheaper material suppliers or reducing expenditure on supervision would be treating symptoms instead of causes and will only make matters worse. But this is what tends to happen.

The proof

A quality manager is unlikely to be able to obtain the cost data needed to prove his case from existing accounting systems. The custodians of financial information tend to guard their territory with ferocity, and any suggestion that the costing system should be enlarged to collect and publish the costs of quality mismanagement will meet opposition from both the accounting and the production functions. What is more, the time taken to establish the cost categories, re-program the computer and set the system in motion is likely to be far too long. But the need is not for precise and comprehensive data. All that is required is enough information to show whether or not an opportunity to make major cost reductions exists and, secondly, where this opportunity is concentrated. For these purposes, no great accuracy is needed. For example, if the cost of poor quality is assessed at 30% of total expenditure, it matters not that this figure may err by as much as 10%, high or low. Even 20% wastage is far too high, and if only half of this can be avoided, the effect on profitability will be substantial.

Figures with an adequate degree of accuracy and reliability can be obtained reasonably inexpensively by means of a pilot study. This may be run on a specific project or work area. As an example, let us consider how a housebuilding company could carry out a study of the quality costs of one particular housing development.

The first step is to select the sample. This must be done with great care. It is usually best to choose a site in the most profitable and best managed part of the organization. The better the management, the more likely they are to understand the purpose of the study and give it their co-operation. Furthermore, if the results show that there are savings to be made on a site which is known to be well run, the conclusion will be drawn that even greater potential benefits will be derived if the experiment is repeated in less well-managed areas.

Having selected the site, it is necessary to devote time, energy and resources to explain the objectives to the site management and to convince them of the benefits which they will receive from the exercise. It may be explained to them that one of the reasons they have been selected is the high regard in which they are held. It is also essential to bring in representatives from higher tiers of management to show that they, too, support the experiment and want it to succeed.

The essence of the pilot study should be to maintain a log of all events and expenses associated with poor quality and to allocate the relevant costs to a small number of cost headings. For a housing development, these may comprise:

Missing or incorrect design information.
Waiting for materials.
Sub-standard materials delivered to site.
Waiting for sub-contractors.
Rectification of sub-contractors' work.
Material wastage.
After-sales remedial works.
Other causes.

The principal burden of collecting costs will inevitably be borne by the site management, but they should be assisted by a member of the quality team whose role would be to make a daily collection of data, to look for costs which might be missed, and to maintain interest. He should ensure that management time spent in dealing with quality problems is picked up and allocated. Some managers spend more than half their time fighting fires and shooting troubles. By the end of, say, one month, it should be apparent whether or not the cost of mismanaging quality on the site is significant. If it is, the next step is to identify the major causes of lost profit so that they may be tackled in order of importance. This may be done by means of a 'Pareto' analysis.

Vilfredo Pareto was an Italian economist whose studies of the distribution of income and wealth enabled him to demonstrate that the major proportion of the world's wealth was held by a minority of the population, while the majority of the population had to share only a minor proportion of the wealth. J.M. Juran found that this principle could be applied to many other activities, including the distribution of quality-related losses (see *Management Breakthrough*, Bibliography). The basis of a Pareto analysis is a tabulation of the cost headings in descending order of magnitude. The usual outcome is that the top 20–30% of items will account for a disproportionately high share of the total cost (perhaps in excess of 70%). Conversely, the lower 70–80% of the table will represent only a minor fraction of the total. Such an analysis enables management to separate the significant few items from the insignificant many. Effort can then be concentrated where it will be most effective.

Table 11.2 is a conjectural example of a Pareto analysis resulting from a pilot study of a housing development. It indicates that two causes of waste stand out above all others as matters in urgent need of management

Table 11.2 Pareto analysis of causes of lost revenue

Item	Percentage of lost revenue
Rectification of sub-contractors' work	35
Missing or incorrect design information	24
Material wastage	11
Waiting for materials	10
Waiting for sub-contractors	7
After-sales remedial works	6
Sub-standard materials	4
Others	3
Total	100

attention. It may be argued that this is the result which most experienced people would expect. But while it remains only a suspicion, nothing will happen. Once identified, the problem areas can be studied more closely and actions taken to eliminate their causes. A successful pilot study can provide the quality system manager with a most valuable weapon. He no longer need be on the defensive, since he can prove his case in a language to which people will listen — the language of cost reduction. Furthermore, he will have concentrated peoples' minds on the subject. In many cases, the act of measuring the costs will have more effect then the actual numbers obtained. To persuade management to accept that these costs are incurred and to participate in their measurement can have wider educational benefits beyond those which may flow from an analysis of the figures. In this case, the process can be more important than the product.

The documents

Having won the propaganda battle, what does our quality system manager do next?

Most construction companies and consultancies already operate quality systems of a kind. Their proud records of successfully completed projects are proof of this. The principal difference between such systems and the more formal arrangements described in this book is that the latter are required to be documented. This documentation presents the greatest obstacle to be overcome. It is likely to be seen by many as unnecessary, bureaucratic and stifling of individual judgment and enterprise. Nevertheless, it is essential. If an organization already has a set of standing instructions or written procedures covering the various functions which affect quality (design, planning, purchasing, sub-contracting, site management, etc.) the first step towards establishing a quality system will already have been made. If not, or if there are gaps, the managers responsible should be required to produce written descriptions of how they expect the activities for which they are responsible to be carried out.

The point has been made several times in this book that responsibility for producing documentation lies with those in charge of the work concerned. It has also been observed that managers and supervisors often show extreme reluctance to commit their requirements to paper. They tend to prevaricate, to wriggle, and to shed their responsibilities on to others. They are likely to explain that they have neither the time nor the resources to prepare written instructions and that, in any case, these are quite unnecessary since all the people in their teams know perfectly well what they have to do already. What is more, it will be said, their departments face imminent reorganization, rationalization or computerization, so there is no point in writing down what is happening to-day since by next month it will all be different. To overcome these, and all the other excuses which will be made, the quality system manager will have to exert constant and unyielding pressure.

Before the quality system manager can exert pressure on managers to produce procedures for work under their control, he has first to establish where the various responsibilities lie. This should be evident from the organization's management structure. Unfortunately, some organizations do not publish their management structure — some even make a virtue of the fact, calling it 'flexibility'. In other cases, the official management hierarchy recognized by senior management may be quite different from the unofficial (but actual) structure developed at lower levels of the organization to accommodate the idiosyncrasies of particular individuals. The quality system manager who succeeds in focusing attention on these matters can already claim a success, since an effective organization structure is a prerequisite of any quality system. Having identified where the real power lies, the quality system manager then has to bring pressure to bear on the individuals concerned. It is almost certain that he will need support from his Chief Executive at this stage, indeed this is likely to be the first test of management's determination to make the system succeed. If the support is not forthcoming, the chances of ultimate success are very slim.

So, the first task is to establish the *status quo* in the form of written procedures produced by, or on behalf of, the functional managers. The quality manager should then assemble and rationalize these to a common format and examine them to see if they contain contradictions, overlaps or gaps. If so, these should be brought to the attention of the managers concerned so that they can make such changes as may be necessary. This, too, can be a most fruitful exercise.

When the procedures have been agreed and issued, it becomes possible to determine the extent to which the quality system they describe already satisfies the requirements of quality system standards. If a decision has been made that the organization will comply with a certain standard, the discrepancies from this standard can be identified and the additional procedures needed to secure compliance can be established. The advantage of this approach is that it enables an organization to identify the additional routines which it needs to introduce to comply with a given system standard. If they are judged beneficial and cost-effective, the managers responsible should willingly introduce them. If they are not judged beneficial, then they can at least be costed so that the true expense of complying with a standard can be assessed satisfactorily.

It is now possible for the quality system manager to complete his documentation by preparing and issuing a quality manual. The content and layout of this document are discussed in some detail in Chapter 5 and do not need further elaboration at this point. It is, however, worth noting that many organizations make the preparation of a quality manual the first step in the establishment of their quality systems. This is only a practicable policy if a reasonably comprehensive set of standing instructions or procedures is already in existence. If this is not so, then the manual will only indicate an intent to create a system, not its existence. It will therefore carry little credibility in the eyes of a potential client or with the organization's own staff.

Quality systems and the control of change

The quality system has now been established and documented, and its performance can be monitored by a programme of audits and management reviews. This does not mean that the quality manager can relax his efforts. The world does not stay the same, and the one aspect of the construction industry which can be predicted with certainty is its continuing unpredictability. A state of change is normality, and when things stop changing it means that either bankruptcy or mortality have taken over. The quality system must therefore be dynamic. It should be an instrument of change and not a means of preserving the *status quo*.

As we move from the twentieth to the twenty-first century, the enterprises which will survive and prosper will be those who learn to perform a difficult double-act. First they must so organize themselves that their product constantly, consistently and without deviation, meets its requirements — they must prevent the unwanted changes which lead to non-conformance. At the same time, they will have continually to transform their processes, systems and structures to cope with ever increasing environmental change. In nature, we find that organisms which are highly specialized for a specific purpose become extinct when the environment changes, and they do not. In the world into which we are moving, organizations will have to develop a capability for stable self-transformation. They will have to learn to manage their own change without at the same time throwing themselves into turmoil.

Many organizations find it difficult to cope with change because they have organization structures established on militaristic lines with vertical lines of authority. Such structures fail to take account of the fact that most of the communications and interactions within teams of people take place along horizontal or lateral lines. Management can find great difficulty in controlling these interactions and may seek to prevent them by dividing the organization into self-contained units who can only communicate with each other through their respective managements. These arrangements not only fragment the processes of the enterprise, they also reinforce the misconception that the primary business functions (marketing, design, production, financial control, etc.) have objectives which are in conflict with each other and with the interests of the purchaser. In pursuit of these false objectives, people can become so proficient in producing substandard goods that it is eventually found necessary to establish specialist quality control departments to identify and reject the dross before it reaches the customer. This is not a good way to run a business. It is tantamount to a drunkard employing a servant to keep him away from the bottle — not only does it not work, it costs money and creates bad feeling.

An effective quality system spans across departmental boundaries. The concept of the internal customer, whose needs must be satisfied, encourages lateral communication and reduces the need for management domination. When management has established a quality culture, people know how they are expected to behave, what is right and what is wrong. They have norms and principles to guide their behaviour and judgement. They are able to make decisions in the best interests of the enterprise when faced with new situations.

But there is a paradox here. Much has been said in this book about the definition of management structures, the preparation of written procedures, formal systems of audit and compliance with system standards. Surely these concepts will lead to a greater emphasis on vertical lines of authority and a strengthened resistance to change? They undoubtedly will if management allows the system to fossilize. This can be prevented from happening by making full use of the upward flow of information generated by the audit process. This arrives on the desks of senior management unfiltered and unadulterated by any intermediate tiers of the hierarchy. It enables them to see the organization as it really is, and not as their subordinates would like them to see it. They are then able to carry out a continual fine-tuning of the management system as it becomes necessary, and not have to wait until it deteriorates to a point at which radical surgery is the only cure.

So we can envisage a future in which people are motivated by being made accountable for satisfying the needs of their internal and external customers in accordance with a company culture which they have come to accept and respect. Managers can then be freed from the need to act as policemen and firefighters and can concentrate on the duty which they, and they alone, can discharge; that is, to refine, adjust and improve management systems to achieve ever higher standards of quality, performance and economy.

This is a vision which may be difficult to sustain when viewed against a backcloth of the rough, tough world of construction. However, we all need a dream of the future, even if it is not immediately achievable. We cannot stand still and we cannot go back. Quality management as it has been described in this book is a goal worth aiming for. It is an objective which can be achieved

APPENDIX A

Glossary of terms

References

1. BS 4778:1987, *Quality Vocabulary, Part 1, International terms.*

   ISO 8402 — 1986, *Quality — Vocabulary.*
2. BS 5882:1980, *Specification for a total quality assurance programme for nuclear installations.*

Defect (1)	The nonfulfilment of intended usage requirements.
Design Review (1)	A formal, documented, comprehensive and systematic examination of a design to evaluate the design requirements and the capability of the design to meet these requirements and to identify problems and propose solutions.
Documentation (2)	Any recorded or pictorial information describing, defining, specifying, reporting or certifying activities, requirements, procedures or results.
Inspection (1)	Activities such as measuring, examining, testing, gauging one or more characteristics of a product or service and comparing these with specified requirements to determine conformity.
Item (2)	An all-inclusive term covering structures, systems, components, parts or materials.
Non-conformity (1)	The nonfulfilment of specified requirements.

Procedure (2) A document that specifies or describes how an activity is to be performed.

Purchaser (2) Any individual or organization who places an order for items or services.

Quality (1) The totality of features and characteristics of a product or service that bear on its ability to satisfy stated or implied needs.

Quality Assurance (1) All those planned and systematic actions necessary to provide adequate confidence that a product or service will satisfy given requirements for quality.

Quality Audit (1) A systematic and independent examination to determine whether quality activities and related results comply with planned arrangements and whether these arrangements are implemented effectively and are suitable to achieve objectives.

Quality Control (1) The operational techniques and activities that are used to fulfil requirements for quality.

Quality Management (1) That aspect of the overall management function that determines and implements the quality policy.

Quality Plan (1) A document setting out the specific quality practices, resources and sequence of activities relevant to a particular product, service, contract or project.

Quality Policy (1) The overall quality intentions and direction of an organization as regards quality, as formally expressed by top management.

Quality Surveillance (1) The continuing monitoring and verification of the status of procedures, methods, conditions, processes, products and services, and analysis of records in relation to stated references to ensure that specified requirements for quality are being met.

Quality System (1) The organizational structure, responsibilities, procedures, processes and resources for implementing quality management.

Quality System Review (1) A formal evaluation by top management of the status and adequacy of the quality system in relation to quality policy and new objectives resulting from changing circumstances.

Repair (2) The process of restoring a non-conforming characteristic to an acceptable condition even though the item may still not conform to the original requirement.

Review (2)
An independent appraisal undertaken by an individual or group competent in the area being considered.

Rework (2)
The process by which an item imade to conform to the original requirement by completion or correction.

Specification (1)
The document that prescribes the requirements with which the product or service has to conform.

Standard (2)
A document approved by a generally recognized body which results from the process of formulating and applying rules for an orderly approach to a specific activity.

Verification (2)
The act of reviewing, inspecting, testing, checking, auditing, or otherwise verifying and documenting whether items, processes, services, or documents conform to specified requirements.

APPENDIX B

Typical quality manual

 *Quality Assurance*	**Alias Construction Ltd** **Bricklayers Road** **London**	Document No: CSI QA-01 Page : Title Date: December 1988

COMPANY QUALITY MANUAL

Company Standing Instruction QA-01

This document does not form part of any contract and is not intended to imply any representation or warranty. The Company reserves the right to amend its procedures from time to time in order to comply with individual contract requirements.

Copy No: *12* **Issued to:** *Procurement Manager*

Issue No.	Amendment to Page 6	Quality Assurance Manager	Date:
4		*R S Jay*	*9 Dec 88*
Issue Date		Chief Executive	Date:
12 Dec 88	REVISION DESCRIPTION	*H T Steel*	*10 Dec 88*

 Quality Assurance	Alias Construction Ltd COMPANY QUALITY MANUAL	Document No: CSI QA - 01 Page: 1 of 16 Date: December 1988

1. CONTROL

1.1 Authority

This document describes the quality system of Alias Construction Ltd and it is issued with my authority.

Departmental and project managers are required to establish and maintain standing instructions and procedures to ensure that work for which they are responsible meets required standards. These instructions or procedures will include appropriate delegation of authority to staff responsible for the achievement and control of quality.

All personnel within the company shall perform their duties in accordance with the requirements of this document and in compliance with company standing instructions and project procedures.

The Company Quality Assurance Manager shall maintain and control the issue of this document, shall verify the implementation of its requirements and shall report to me on the status and effectiveness of its implementation.

Any disputes concerning the requirements of this document which cannot be resolved by the Quality Assurance Manager shall be referred to me.

H T Steel

Chairman and Chief Executive 4th January 1986

 *Quality Assurance*	**Alias Construction Ltd** **COMPANY QUALITY MANUAL**	Document No: CSI QA-01 Page: 2 of 16 Date: December 1988

1. CONTROL (continued)

1.2 **Table of Contents**

Section No Description

<table>
<tr><td rowspan="3">
Quality Assurance</td><td rowspan="3">Alias Construction Ltd
COMPANY QUALITY MANUAL</td><td>Document No:
CSI QA-01</td></tr>
<tr><td>Page:
3 of 16</td></tr>
<tr><td>Date:
December 1988</td></tr>
</table>

1. CONTROL (continued)

1.3 Change Control

This Manual is revised periodically to keep it up to date. The issue number and date of the current edition are indicated on the title page. Subsequent pages carry the issue date only. Latest revisions are marked in the text with a vertical line in the right hand margin labelled with the issue number in a triangle.

Controlled copies of the Manual are numbered and a register is kept of recipients so that none will be left out when new issues are made. Holders of controlled copies are required to destroy out-of-date copies.

Uncontrolled copies of the Manual are stamped "Not subject to change control' on the title page.

1.4 Schedule of Manual Holders

Title	Controlled Copy No
Chairman	01
Operations Director	02
Commercial Director	03
Administration Manager	04
Quality Assurance Manager	05
Business Development Manager	06
Group Quality Assurance Manager	07
Engineering Manager	08
Estimating Manager	09
Construction Manager	10
Plant and Transport Manager	11
Procurement Manager	12
Chief Quantity Surveyor	13
Stores Manager	14
Chief Safety Officer	15
Training Manager	16
Project Managers	50/1, 50/2, 50/3 etc
Project Quality Assurance Engineers	51/1, 51/2, 51/3 etc
Other Listed Holders	52/1, 52/2, 52/3 etc

A full schedule of holders is maintained by the Company Quality Assurance Manager.

Quality Assurance

Alias Construction Ltd
COMPANY QUALITY MANUAL

2. CORPORATE POLICY

2.1 Objective

The corporate objective of the Alias Group is to maintain and improve upon its position as a leading construction contractor and builder of houses. To achieve this objective, operating companies within the Group are required to devise and implement management systems which will:

- Ensure that the requirements of clients and customers are satisfied.
- Maintain the safety, skills, welfare and morale of the work-force.
- Ensure the generation of sufficient profit to provide an adequate return on capital employed.

2.2 Quality Systems

The management systems maintained by operating companies to ensure client and customer satisfaction are known as quality systems. Because of the wide range of activities undertaken by the Group, it is not practicable for there to be a single quality system common to all subsidiary companies. Instead, each is required to develop and document a system suitable for its own activities, including organization, delegation of authority, and arrangements for audit and surveillance of the company's own activities and those of its suppliers and sub-contractors.

Companies which contain a number of specialist divisions for which it is not practicable to establish a single quality system may operate separate systems for each speciality, subject to approval and audit at company level.

Associate companies for which Alias assumes a management responsibility have the option of operating within the quality system of their Alias parent or adopting a quality system appropriate to their work and circumstances and approved by their board of directors.

<table>
<tr><td rowspan="3">
Quality Assurance</td><td rowspan="3">Alias Construction Ltd
COMPANY QUALITY MANUAL</td><td>Document No:
CSI QA-01</td></tr>
<tr><td>Page:
5 of 16</td></tr>
<tr><td>Date:
December 1988</td></tr>
</table>

2. CORPORATE POLICY (continued)

2.3 Responsibilities of Management

All staff charged with managerial responsibility are required to ensure:

- that work carried out under their control complies with specification. This includes work undertaken by sub-contractors.
- that staff under their jurisdiction are familiar with the requirements of the quality system relevant to their work, that they have access to all applicable company and project procedures and that they have read and understood these procedures.
- that staff assigned to a project or site are adequately qualified and experienced in their relevant technical discipline to perform their duties in a satisfactory manner.

2.4 Group Quality Assurance

Reporting relationships in respect of quality assurance at Group level are illustrated diagrammatically on Exhibit "A' of this Manual.

The Group Technical Director reports to the Group Chief Executive on the implementation and effectiveness of company quality systems.

The Group Quality Assurance Manager is responsible to the Group Technical Director for the following functions:

- Advice to operating companies on quality systems and documentation.
- Co-ordination of recruitment and training of quality assurance staff.
- Presentations to clients.
- Preparation of periodic reports to the Group Technical Director for board presentation.
- Auditing of company quality systems.

 Quality Assurance	Alias Construction Ltd COMPANY QUALITY MANUAL	Document No: CSI QA-01 Page: 6 of 16 Date: December 1988

2. CORPORATE POLICY (continued)

2.4 **Group Quality Assurance** (continued)

Quality systems of operating companies are required to be reviewed annually by the director or senior manager responsible for their operation. Reports of their reviews are submitted to company chief executives and to the Group Quality Assurance Manager by 30 November each year.

Group audits of operating companies may be instigated by the Group Technical Director on the recommendation of the Group Quality Assurance Manager or they may be requested by company managements. Auditors for group audits are selected by the Group Quality Assurance Manager.

A report based on the above reviews and audits is submitted by the Group Quality Assurance Manager to the Group Technical Director on 1 January each year.

 *Quality Assurance*	**Alias Construction Ltd** **COMPANY QUALITY MANUAL**	Document No: CSI QA-01
		Page: 7 of 16
		Date: December 1988

3. COMPANY ORGANIZATION

3.1 Description

The principal activities of Alias Construction Ltd are building and civil engineering. It undertakes medium to large contracts, mainly in the United Kingdom.

The position of Alias Construction Ltd within the Alias Group is illustrated on Exhibit "B'.

3.2 Management Structure

The management structure of Alias Construction Ltd is illustrated on Exhibit "C'. Two directors and three managers report to the Chief Executive. Their roles are as follows:

The OPERATIONS DIRECTOR is responsible for the execution of projects. Four departmental managers answer to him for the following functions:

Construction
Engineering
Estimating
Plant and Transport

The control of quality to ensure compliance with specification is the responsibility of line management answerable to the Operations Director. It follows that the preparation of work instructions, training of personnel, the supervision of sub-contractors and suppliers, and the execution of testing programmes are all line management responsibilities.

The COMMERCIAL DIRECTOR controls contractual and commercial aspects of contracts and sub-contracts and assigns commercial personnel to projects. Departmental Managers responsible for the following functions report to him:

Financial Reporting
Quantity Surveying
Accounts
Procurement

The BUSINESS DEVELOPMENT MANAGER is responsible for marketing the company's services, for establishing and maintaining contacts with potential clients and for preparing pre-qualification submissions.

Quality Assurance

Alias Construction Ltd
COMPANY QUALITY MANUAL

Document No: CSI QA-01
Page: 8 of 16
Date: December 1988

3. **COMPANY ORGANIZATION** (continued)

3.2 **Management Structure** (continued)

The ADMINISTRATION MANAGER is responsible for the functions of office and site administration, including time and pay, storekeeping and site costing.

The QUALITY ASSURANCE MANAGER is responsible for maintaining the effectiveness of the company quality system.

3.3 **Management of Quality Assurance**

The Company Quality Assurance Manager has line responsibility to the Chief Executive and acts independently of departmental and project managers. He has functional responsibility for quality assurance engineers assigned to projects, although they report on a day-to-day basis to their respective project managers. The Quality Assurance Manager's duties include:

- Managing the Quality Assurance Department.
- Acting as custodian of Company Standing Instructions.
- Advising on the preparation of project quality plans.
- Auditing the implementation of the Company quality system.
- Ensuring that system defects revealed by audits are resolved by the appropriate manager.
- Conducting or arranging for audits on vendors and suppliers in collaboration with the Procurement Manager.
- Representing Alias Construction during quality audits by clients or certifying bodies.
- Carrying out annual reviews of the system in accordance with Group instructions and reporting the results to the Chief Executive.

 *Quality Assurance*	**Alias Construction Ltd** **COMPANY QUALITY MANUAL**	Document No: **CSI QA-01**
		Page: **9 of 16**
		Date: **December 1988**

4. COMPANY STANDING INSTRUCTIONS

4.1 General

Standard procedures to be used for carrying out the Company's business are prescribed in Company Standing Instructions (CSIs).

CSIs are prepared and updated by the appropriate departmental heads in accordance with CSI QA-02 "Preparation and Administration of Instructions and Procedures'. The Company Quality Assurance Manager is responsible for the controlled issue of CSIs and for standardizing their format.

4.2 Compliance with Quality System Standards

Exhibit "D' lists the CSIs which address the requirements of BS 5750 : Part 1 and BS 5882, and gives cross reference to the relevant clauses.

4.3 Scope of CSIs

Each CSI includes a statement of scope. The scopes of the CSIs listed in Exhibit "D' are reproduced below:

CSI QA-02 : Preparation and Administration of Instructions and Procedures

Scope : This instruction describes the preparation, format, numbering, revision and administration of standing instructions and project procedures and their distribution as part of the Company's quality system.

This instruction serves as an example of how instructions and procedures should be presented.

The Company Quality Manual, although issued and administered as a company standing instruction, is excluded from the requirements of this instruction.

CSI QA-03 : Quality Audits

Scope : This instruction relates to internal system audits and external audits of suppliers and sub-contractors. It identifies criteria for the selection of auditors and prescribes procedures for audit planning, preparation, performance, follow-up and close-out.

 Quality Assurance	Alias Construction Ltd COMPANY QUALITY MANUAL	Document No: CSI QA-01 Page: 10 of 16 Date: December 1988

4. COMPANY STANDING INSTRUCTIONS (continued)

4.2 Scope and Compliance with Quality System Standards (contd)

CSI QA-04 : Corrective Action

Scope : This instruction describes procedures to be followed to investigate the causes of non-conforming work and to determine the actions necessary to prevent repetition.

CSI QA-05 : Quality Assurance Records

Scope This instruction assigns responsibilities and specifies procedures for the collection, classification, storage and retrieval of records.

CSI OPS-01: Project Organization

Scope : This instruction describes procedures for the initiation, execution and close-down of projects. It identifies the principal responsibilities attached to senior management posts. It defines interfaces between departments and establishes lines of communication and reporting procedures.

CSI OPS-02: Construction Planning

Scope : This instruction describes procedures for the review of contract requirements and for the development of plans for their achievement.

CSI OPS-03: Quality Plans

Scope : This instruction assigns responsibilities and defines procedures for the preparation, issue, review and updating of project quality plans.

CSI OPS-04: Work Instructions

Scope : This instruction assigns responsibilities and defines procedures for the identification, preparation and issue of instructions for the control of work. It requires that such instructions should include construction methods, sampling procedures, criteria for workmanship, and the disposition of non-conforming items, as appropriate.

Quality Assurance

Alias Construction Ltd

COMPANY QUALITY MANUAL

4. COMPANY STANDING INSTRUCTIONS (continued)

4.2 Scope and Compliance with Quality System Standards (contd)

CSI OPS-07: Design of Permanent Works

Scope : This instruction assigns responsibilities and specifies procedures for the commissioning and control of external designers and consultants engaged by the company.

CSI OPS-08: Design and Construction of Temporary Works

Scope : This instruction assigns responsibilities and defines procedures for controlling the design, erection, use and removal of temporary works.

CSI OPS-09: Documentation and Change Control

Scope : This instruction assigns responsibilities and specifies procedures for controlling the receipt, storage and issue of drawings, specifications and other documents. It includes procedures for controlling changes to documents and for preventing the use of obsolete versions.

CSI OPS-12: Inspection and Test Equipment

Scope : This instruction defines the actions to be taken to ensure that inspection, measuring, test and control equipment is adequately maintained and calibrated.

CSI OPS-14: Inspection and Test Plans

Scope : This instruction establishes procedures for the preparation of inspection and test plans. It includes samples of all standard forms used for recording the results of inspections and tests and gives directions on their use.

Quality Assurance

Alias Construction Ltd

COMPANY QUALITY MANUAL

Document No: CSI QA-01

Date: December 1988

4. COMPANY STANDING INSTRUCTIONS (continued)

4.2 Scope and Compliance with Quality System Standards (contd)

CSI COM-03: Purchasing

Scope : This instruction assigns responsibilities and defines procedures to be used for the procurement of goods and services to ensure their compliance with specification and on-time delivery.

CSI COM-05: Selection and Control of Sub-contractors

Scope : This instruction assigns responsibilities and prescribes procedures to be used for the selection, appointment, control and payment of sub-contractors.

CSI COM-07: Administration of Contract Variations

Scope : This instruction relates both to main contracts with clients and to sub-contracts for the supply of materials and services. It assigns responsibilities and specifies procedures for recording, evaluating, communicating and implementing contractual variations.

CSI COM-10: Commercial Records

Scope : This instruction assigns responsibilities and describes procedures for the collection, classification, storage and retrieval of records required for commercial purposes.

CSI ADM-06: Material Receipt, Storage and Issue

Scope : This instruction assigns responsibilities and describes procedures for the receipt, handling, storage and issue of materials.

4. COMPANY STANDING INSTRUCTIONS (continued)

4.2 **Scope and Compliance with Quality System Standards** (contd)

CSI ADM-08: Training Records

Scope : This instruction assigns responsibilities and specifies procedures for the compilation, recording and up-dating of staff training records.

 *Quality Assurance*	**Alias Construction Ltd** **COMPANY QUALITY MANUAL**	Document No: CSI QA-01
		Page: 14 of 16
		Date: December 1988

5. PROJECT QUALITY ASSURANCE

5.1 Staff Responsibilities

The Company Quality Assurance Manager assesses the needs of each new project and assigns quality assurance staff either on a full-time or visiting basis. All major projects have full-time project quality assurance engineers.

Full-time project quality assurance engineers operate exclusively within their project teams and report to their Project Managers, but they retain a functional link to the Company Quality Assurance Manager and are responsible to him for implementing the company quality system on their projects.

Visiting quality assurance engineers are answerable to the Company Quality Assurance Manager.

The duties of project quality assurance engineers include:

- Preparing and updating project quality plans in collaboration with project staff.
- Attending quality plan review meetings.
- Carrying out internal and supplier audits in accordance with the quality plan.
- Following-up corrective actions resulting from audits.
- Commissioning and overseeing specialist on-site inspection services.
- Co-ordinating the off-site inspection of purchased materials.
- Co-ordinating the collation and storage of record documentation.
- Responding to internal and external audits of the project.
- Liaison with client's representatives to establish procedures which will effectively discharge any contractual obligations in respect of inspection, documentation or records.

Project quality assurance engineers are **not** responsible for making technical assessments of construction work or for making construction decisions. These functions are delegated to technical specialists from the Construction or Engineering Departments.

 *Quality Assurance*	**Alias Construction Ltd** **COMPANY QUALITY MANUAL**	Document No: CSI QA-01
		Page: 15 of 16
		Date: December 1988

5. **PROJECT QUALITY ASSURANCE** (continued)

5.2 **Project Procedures**

Project Procedures (PPs) are documents which specify standard methods of operation unique to a particular project. They are prepared to cover subjects not catered for in existing CSIs or when the use of a particular CSI would be detrimental to the interests of a particular project as, for example, when a client specifies or requires an alternative procedure.

For further information on Project Procedures, see CSI QA-02 "Preparation and Administration of Instructions and Procedures'.

5.3 **Project Quality Plans**

On award of contract, the designated project manager and his chief or senior engineer, in collaboration with the project quality assurance engineer, is required to prepare the project quality plan. The plan will be documented to meet specific project requirements, and will be subject to approval by the client if this is a contractual requirement.

Project quality plans will typically include:

(a) A schedule of drawings and other documentation specifying the work to be done.

(b) A description of the project organization and allocation of responsibilities.

(c) A schedule of Company Standing Instructions, project procedures and work instructions relevant to the project.

(d) A schedule of inspection and test plans.

(e) A timetable of quality audits.

(f) The arrangements for review and update of the quality plan.

(g) Other sections as required under the contract.

For further information on project quality plans, see CSI OPS-03 "Quality Plans'.

 *Quality Assurance*	Alias Construction Ltd COMPANY QUALITY MANUAL	Document No: CSI QA-01 Page: 16 of 16 Date: December 1988

6. EXHIBITS

A : Reporting Relationships : Group Quality Assurance

B : Group Organizational Structure

C : Company Management Structure

D : Company Standing Instructions.

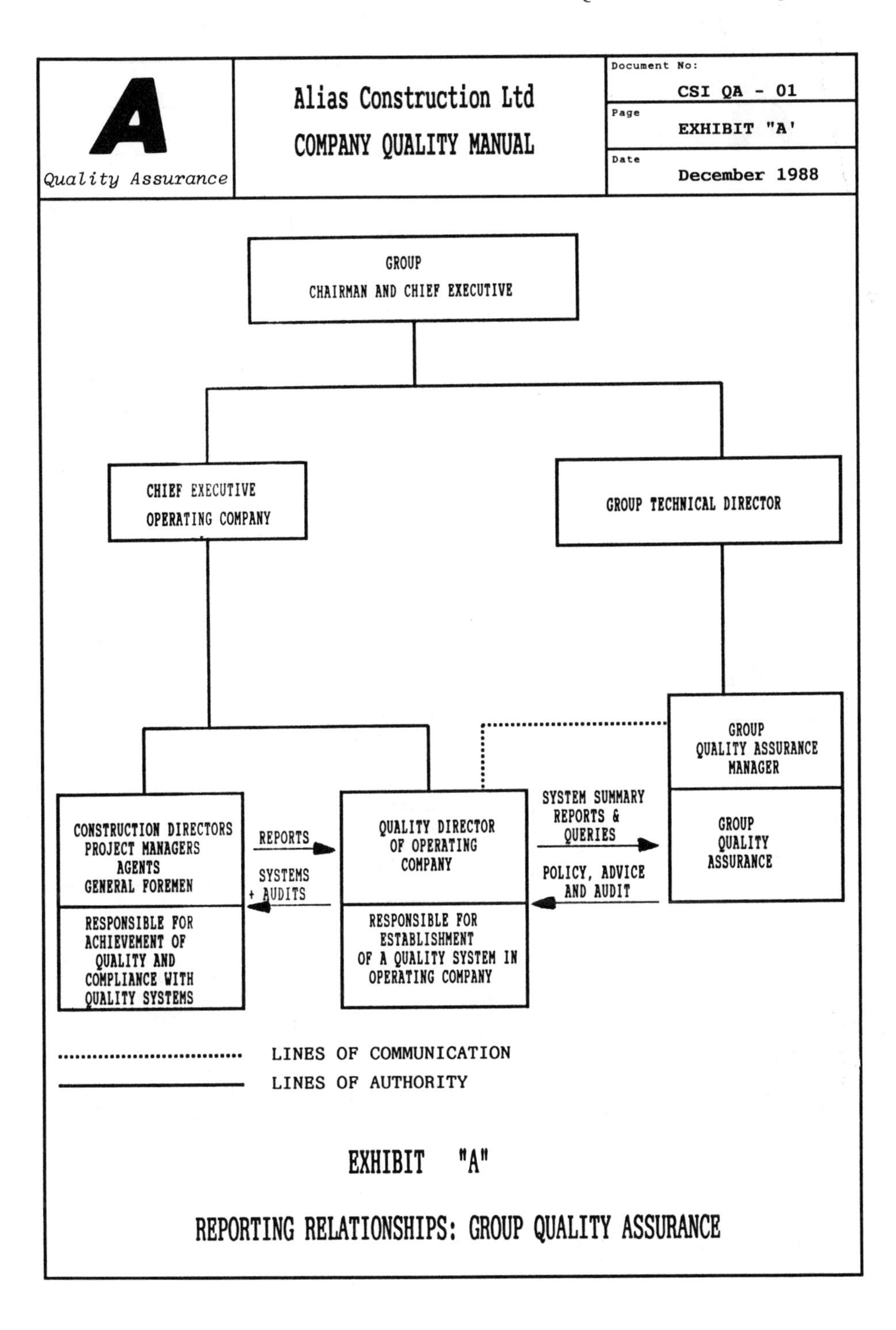

EXHIBIT "A"

REPORTING RELATIONSHIPS: GROUP QUALITY ASSURANCE

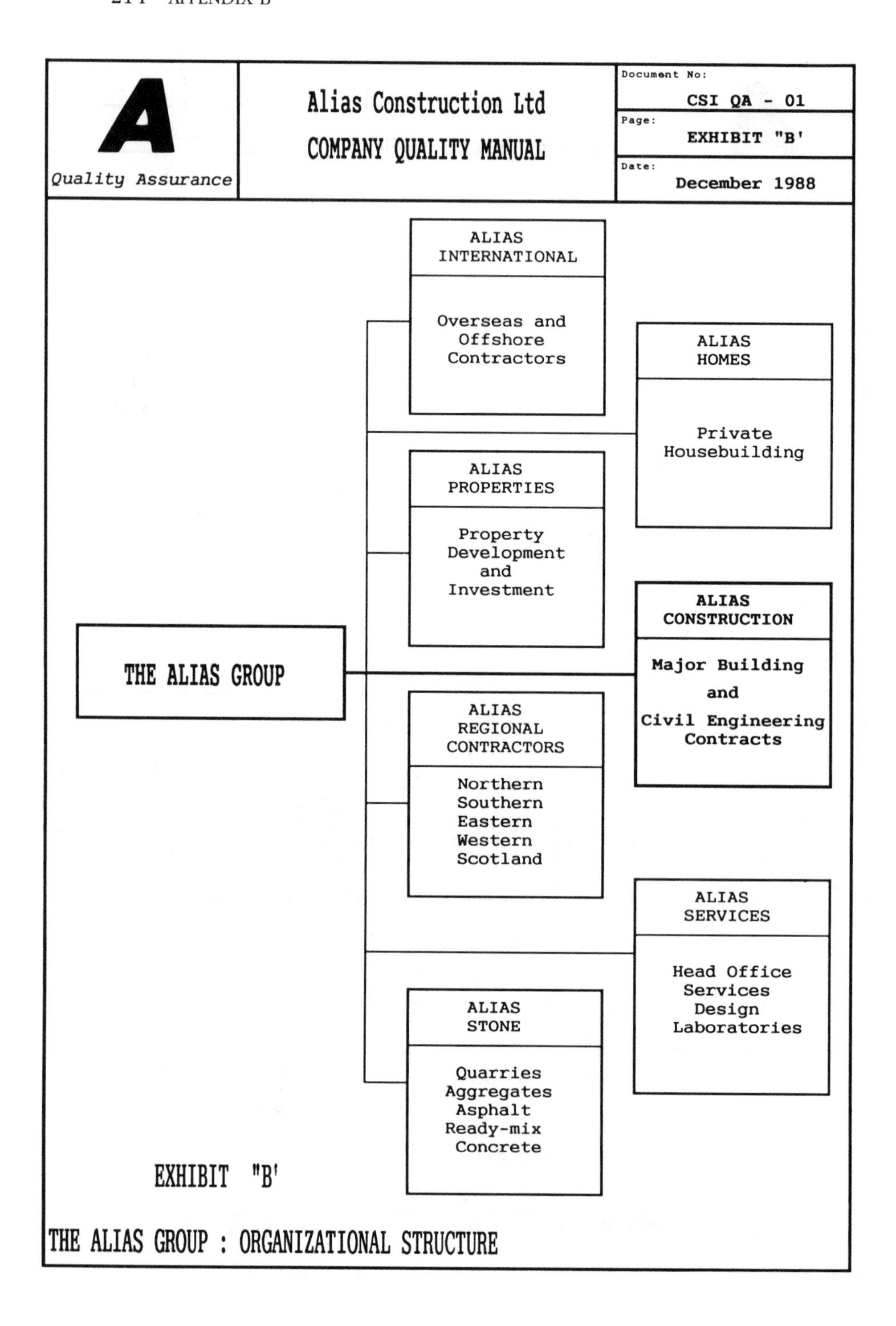

THE ALIAS GROUP : ORGANIZATIONAL STRUCTURE

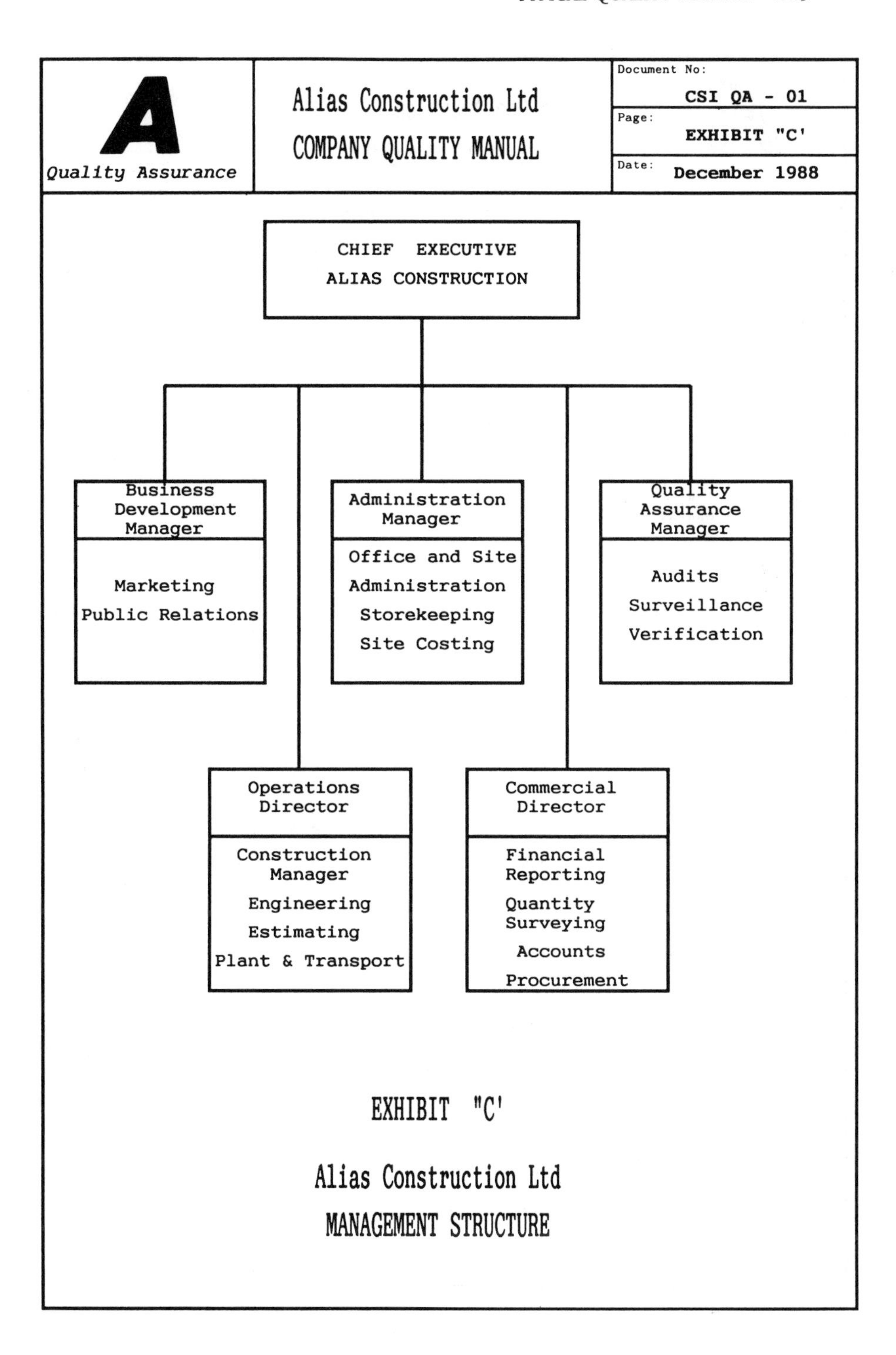

EXHIBIT "C'

Alias Construction Ltd
MANAGEMENT STRUCTURE

Alias Construction Ltd
COMPANY QUALITY MANUAL

Document No: CSI QA-01
Page: EXHIBIT "D'
Date: December 1987

No.	Company Standing Instruction	BS 5750: Part 1 Clause	BS 5882 Clause
	Responsibility of Quality Assurance Manager		
QA-01	Company Quality Manual	4.1, 4.2, 4.3	1, 2
QA-02	Preparation and Administration of Instructions and Procedures	4.2	5
QA-03	Quality Audits	4.17	18
QA-04	Corrective Action	4.14	17
	Responsibility of Operations Director		
OPS-01	Project Organization	4.1, 4.3	2
OPS-02	Construction Planning	4.3, 4.9	-
OPS-03	Quality Plans	4.2, 4.9	1
OPS-04	Work Instructions	4.9, 4.13, 4.20	5, 8, 15
OPS-07	Design of Permanent Works	4.4, 4.20	3
OPS-08	Design and Construction of Temporary Works	4.4	3
OPS-09	Documentation and Change Control	4.5	6
OPS-12	Inspection and Test Equipment	4.11	12
OPS-14	Inspection and Test Plans	4.8, 4.10, 4.12, 4.16	9, 10, 11, 14
	Responsibility of Commercial Director		
COM-03	Purchasing	4.6	4.7
COM-05	Selection and Control of Sub-contractors	4.6	4.7
COM-07	Administration of Contract Variations	4.5	6
COM-10	Commercial Records	4.16	17
	Responsibility of Administration Manager		
ADM-06	Material Receipt, Storage and Issue	4.7, 4.15	13
ADM-08	Training Records	4.18	2

EXHIBIT "D'

COMPANY STANDING INSTRUCTIONS AND

QUALITY SYSTEM STANDARDS

APPENDIX C

Typical procedure

 *Quality Assurance*	**Alias Construction Ltd** **Bricklayers Road** **London**	Document No: CSI QA - 02 Page: Title Date: June 1988

PREPARATION AND ADMINISTRATION
OF INSTRUCTIONS AND PROCEDURES

Company Standing Instruction QA-02

This document is the property of ALIAS CONSTRUCTION LTD, and its issue is controlled. The information contained herein may not be disclosed, in whole or in part, either verbally or in writing, without prior consent of the Company in writing.

This document does not form part of any contract and is not intended to imply any representation or warranty. The Company reserves the right to amend its procedures from time to time in order to comply with individual contract requirements.

Issue No.		Prepared by: *H Beam*	Date: *1 June 1988*
2	**Amendment to Page 5**	Quality Assurance Manager *R S Jay*	Date: *5 June 1988*
Issue Date		Department/Project Manager	Date:
12 June 1988	REVISION DESCRIPTION	Client	Date:

 Quality Assurance	Alias Construction Ltd **PREPARATION AND ADMINISTRATION OF INSTRUCTIONS AND PROCEDURES**	Document No: CSI QA-02 Page: 1 of 7 Date: June 1988

TABLE OF CONTENTS

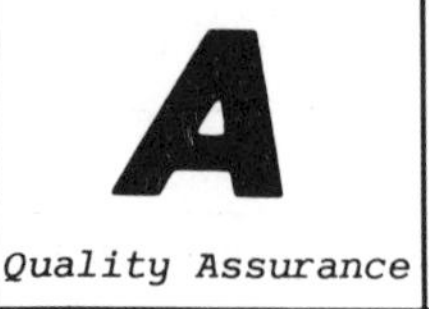 *Quality Assurance*	**Alias Construction Ltd** **PREPARATION AND ADMINISTRATION OF INSTRUCTIONS AND PROCEDURES**	Document No: CSI QA-02
		Page: 2 of 7
		Date: June 1988

1. INTRODUCTION

Alias Construction Ltd is hereinafter referred to as the Company. The Company's policies on the preparation and use of company standing instructions and other procedure documents are stated in the Company Quality Manual CSI QA-01.

2. SCOPE

This instruction describes the preparation, format, numbering, revision and administration of standing instructions and project procedures and their distribution as part of the Company's quality system.

This instruction serves as an example of how instructions and procedures should be presented.

The Company Quality Manual, although issued and administered as a company standing instruction, is excluded from the requirements of this instruction.

3. PERSONNEL RESPONSIBLE

Title : Company Quality Assurance Manager

Responsibilities: To administer company standing instructions on behalf of departmental managers in accordance with the Company Quality Manual and this procedure.

Title : Departmental directors/managers

Responsibilities: To establish and maintain company standing instructions which describe and document the activities of their functions. To review and approve project procedures.

Title : Project Managers

Responsibilities: To establish and maintain approved project procedures as listed in the Project Quality Plan (see CSI OPS-02).

To obtain client approval when necessary.

 *Quality Assurance*	**Alias Construction Ltd** **PREPARATION AND ADMINISTRATION OF INSTRUCTIONS AND PROCEDURES**	Document No: CSI QA-02
		Page: 3 of 7
		Date: June 1988

4. PROCEDURE

4.1 **Company Standing Instructions**

Company Standing Instructions (CSIs) prescribe the standard procedures to be used for carrying out the company's business.

Departmental directors and managers are required to produce draft instructions in which the activities for which they are directly or functionally responsible are fully described, specified and documented. They may delegate the writing of instruction drafts to responsible persons in their departments.

Draft instructions shall be laid out in accordance with the requirements of this instruction and must be approved by departmental heads before submission to the Company Quality Assurance Manager for review and comment prior to issue by his department.

The Company Quality Assurance Manager is responsible for the presentation, format and controlled issue of CSIs. The contents are the responsibility of departmental heads.

All necessary revisions and up-dating of CSIs are to follow the above procedure.

4.2 **Project Procedures**

Project Procedures (PPs) are instructions produced for project use when the strict application of CSIs can be shown to be inappropriate (for example, when a client requires or specifies an alternative procedure) or when the particular subject is not adequately addressed in existing CSIs.

Project Procedures will be prepared by project staff for review and approval by Company departmental heads. The Project Manager is responsible for obtaining client approval if necessary and for distribution of copies thereafter.

PPs will be produced as amendments or supplements to existing CSIs. They will take the form of a schedule of amendments and supplements, with cross reference to amended clauses and a reproduction of the full text of each changed clause or supplementary clause as appropriate.

PPs submitted to clients for approval require a full copy of the current issue of the appropriate CSI to be appended to the PP for immediate reference.

 *Quality Assurance*	**Alias Construction Ltd** **PREPARATION AND ADMINISTRATION OF INSTRUCTIONS AND PROCEDURES**	Document No: CSI QA-02
		Page: 4 of 7
		Date: June 1988

4. PROCEDURE (continued)

4.2 **Project Procedures** (continued)

The Company Quality Assurance Manager will receive copies of all PPs and is required to note their contents for possible inclusion in appropriate CSIs at the next review and subsequent revision.

4.3 **Format**

CSIs shall follow the format of this document. Paragraph numbering and layout will be as demonstrated. The title page will be identical to that at the front of this Instruction.

Printing will be on one side of the paper only.

4.4 **Contents**

CSIs shall include, but not be limited to, the following sections:

(a) TABLE OF CONTENTS.

(b) INTRODUCTION. This will state the purpose and background of the Instruction.

(c) SCOPE. This will describe the subject matter of the instruction.

(d) PERSONNEL RESPONSIBLE. This will identify the persons carrying responsibility for the activities described in the instruction.

(e) PROCEDURE. This will state how the activities are to be carried out, if necessary on a step-by-step basis, and will allocate responsibility for each activity.

(f) REFERENCES. This will list relevant standards and other documents to which reference should be made.

 *Quality Assurance*	**Alias Construction Ltd** **PREPARATION AND ADMINISTRATION OF INSTRUCTIONS AND PROCEDURES**	Document No: CSI QA-02 Page: 5 of 7 Date: June 1988

4. PROCEDURES (continued)

4.4 Contents (continued)

(g) EXHIBITS. These may include the following:

Organization charts and diagrams

Flow charts

Schedules

Company pro-formae

Other company documents relevant to the procedure

Exhibits shall be included in the procedure immediately after a list of all exhibits as follows:

EXHIBIT 'A' - (Title)

EXHIBIT 'B' - (Title)

4.5 Numbering of Instructions and Procedures

CSIs shall be numbered as follows:

(a) The letters CSI indicating Company Standing Instruction.

(b) A group of letters to denote the department responsible for originating the procedures it describes. These are:

Symbol	Department
ADM	Administration
COM	Commercial
OPS	Operations
QA	Quality Assurance

Quality Assurance

Alias Construction Ltd

PREPARATION AND ADMINISTRATION OF INSTRUCTIONS AND PROCEDURES

4. **PROCEDURE** (continued)

4.5 **Numbering of Instructions and Procedures** (continued)

(c) A sequence number made up of two digits, e.g. "03".

For example, the CSI issued by Operations Department in respect of Inspection and Test Equipment is numbered:

CSI OPS-12

PPs issued as amendments or supplements to CSIs will bear the original CSI number, but this will be prefixed by the appropriate Project Job Number. For example, a PP covering Inspection and Test Equipment on Job No 15/306 might be numbered:

15/306 PP OPS-12

4.6 **Numbering of Pages**

The title page shall not have a number.

Continuation pages will be numbered consecutively, commencing at page 1. All pages will show the total number of pages. Exhibits will not be included in the page numbering.

4.7 **Revisions**

Revisions to CSIs and PPs will be identified by the Issue Number and date shown on the title page and by the date on each succeeding page. Revisions will be issued as complete instructions or procedures.

The first and subsequent issues of instructions or procedures for APPROVAL will be identified as:

ISSUE A
ISSUE B
ISSUE C, etc.

The first and subsequent issues of APPROVED instructions and procedures will be identified as:

ISSUE 0
ISSUE 1
ISSUE 2, etc.

 *Quality Assurance*	**Alias Construction Ltd** **PREPARATION AND ADMINISTRATION OF INSTRUCTIONS AND PROCEDURES**	Document No: CSI QA-02
		Page: 7 of 7
		Date: June 1988

4. **PROCEDURE** (continued)

4.7 **Revisions** (continued)

Details of revisions listing the pages which have been amended will be entered on the title page. Revisions will be marked in the text with a vertical line in the right hand margin with the issue number in a triangle. With each subsequent issue, the previous revision indicators will be deleted, and only the current revisions will be indicated.

4.8 **Distribution**

The Company Quality Assurance Manager will maintain a register of copy-holders of CSIs and will issue 'controlled' copies to them. Issues will be accompanied by Document Distribution Schedules (see CSI OPS-09).

Project Managers, or Project Quality Assurance Engineers on their behalf, will maintain a register of copy-holders of PPs and will issue 'controlled' copies to them, accompanied by Document Distribution Schedules.

Recipients of new issues of CSIs and PPs are required to remove and destroy superseded copies. They are also responsible for ensuring that appropriate subordinate personnel are made aware of instructions and procedures which affect their work.

5. **REFERENCES**

BS 5750 : Part 1	Quality systems : specification for design/development, production, installation and servicing.
BS 5750 : Part 2	Quality systems : specification for production and installation.
BS 5750 : Part 3	Quality systems : specification for final inspection and test.
CSI QA-01	Company Quality Manual
CSI OPS-09	Documentation and Change Control

APPENDIX D

Typical quality plan

<table>
<tr><td>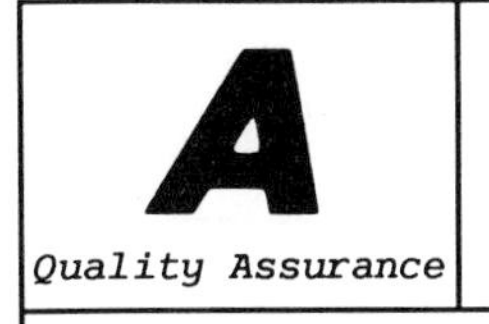
Quality Assurance</td><td>Alias Construction Ltd
Bricklayers Road
London</td><td>Document No: ASTW-QP-01
Page: Title
Date: June 1988</td></tr>
</table>

QUALITY PLAN

ARCADIA SUPERMARKET

TUNBRIDGE WELLS

<table>
<tr><td>Issue No.
0</td><td rowspan="2">
REVISION DESCRIPTION</td><td>Prepared by:
H Beam</td><td>Date:
1 June 1988</td></tr>
<tr><td>Issue Date
12 June 1988</td><td>Quality Assurance Manager
R S Jay</td><td>Date:
5 June 1988</td></tr>
<tr><td></td><td></td><td>Department/Project Manager
P Murphy</td><td>Date:
7 June 1988</td></tr>
<tr><td></td><td></td><td>Client</td><td>Date:</td></tr>
</table>

 *Quality Assurance*	Arcadia Supermarket Tunbridge Wells QUALITY PLAN	Document No: **ASTW - QP - 01** Page: 1 of 4 Date: **June 1988**

1. **GENERAL**

This quality plan is prepared in accordance with CSI OPS-03 'Quality Plan'.

The plan shall comprise this document together with the following:

a) Company Quality Manual CSI QA-01 Issue 4.

b) Company Standing Instructions (subject to any amendments listed in Section 4 hereunder).

c) The drawings and specifications (see Section 2).

d) Project Procedures (see Section 4).

e) Work Instructions (see Section 5).

f) Inspection, test and audit plans (see Section 6).

2. **QUALITY OBJECTIVES**

The project will be constructed in accordance with the following documents:

a) Contract Specification ASTW-01 Pages 1 to 57.

b) Architect's drawings numbered ASTW-A-01 to ASTW-A-56.

c) Structural Engineer's drawings numbered ASTW-S-01 to ASTW-S-25.

d) Steelwork sub-contractor's drawings numbered ASTW-SW-01 to ASTW-SW-15.

Master copies of the above documents are held in the Project Engineer's office, and he is responsible for keeping them up-to-date and in clean condition. He will control the receipt and issue of drawings and specifications in accordance with CSI OPS-09.

3. **ALLOCATION OF RESPONSIBILITIES**

3.1 **Duties**

The Project Manager is responsible for issuing the quality plan and for ensuring that all work complies with the drawings and specifications.

The Project Engineer is responsible for verifying that materials and workmanship comply with specification before payments are made. He will be responsible for collecting and storing quality records. He will report departures from this quality plan to the Project Manager for his resolution.

 Quality Assurance	Arcadia Supermarket Tunbridge Wells QUALITY PLAN	Document No: ASTW - QP - 01
		Page: 2 of 4
		Date: June 1988

3. ALLOCATION OF RESPONSIBILITIES (contd)

The Regional Quality Engineer will prepare revisions of the plan, as necessary, for approval and issue by the Project Manager. (See Section 7) He will audit its implementation in accordance with the programme in Section 6.

The Project Buyer is responsible for ensuring that all orders for purchased materials are in compliance with specifications.

3.2 Organization

The site organization structure will be as follows:

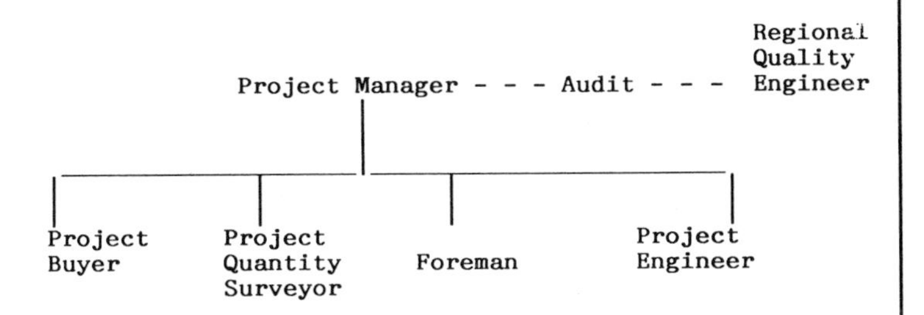

4. PROJECT PROCEDURES

The procedure specified in the contract for processing contract variations conflicts with the requirements of CSI COM- 07 "Administration of Contract Variations'. The project quantity surveyor will prepare an amendment to CSI COM-07 incorporating the necessary changes. After approval by the Commercial Director this will be issued as Project Procedure ASTW PP-COM-07.

Quality Assurance	Arcadia Supermarket Tunbridge Wells QUALITY PLAN	Document No: ASTW - QP - 01
		Page: 3 of 4
		Date: June 1988

5. WORK INSTRUCTIONS

The following sections of the work are unusual or complex and will require the issue of written work instructions:

5.1 Basement Excavation

Work Instruction No.1 to be prepared by Project Engineer and General Foreman for approval by Project Manager.
Target date for issue: Week 4.

5.2 Concreting

Work Instruction No.2 covering transportation, placement and compaction of concrete to be prepared by Project Engineer for approval by Project manager.
Target date for issue: Week 4.

5.3 Structural Steelwork Erection

Work Instruction No.3 to be produced by sub-contractor for approval by Project Manager. Buyer will make this a condition of the sub-contract. Target date for issue: Week 10.

5.4 Cladding

Use instruction entitled "Profiled Sheet Metal Roofing and Cladding' issued by NFRC.

5.5 Mastic Roof

Project Engineer to request Alias Mastic Consultant to prepare Work Instruction No.4. Target Date for issue: Week 12.

6. INSPECTION, TESTING AND AUDITS

6.1 Inspection and Test Plans

The Project Engineer will compile and issue inspection and test plans for the following operations:

Excavation & fill	(ASTW-ITP-1)	Target date: Week 4
Concreting	(ASTW-ITP-2)	Target date: Week 6
Brickwork	(ASTW-ITP-3)	Target date: Week 6

The following sub-contractors will submit inspection and test plans to the Project Engineer for his approval. Buyer will make this a condition of contract.

Piling	(ASTW-ITP-SC1)	Target date: Week 4
Structural Steelwork	(ASTW-ITP-SC2)	Target date: Week 10

Quality Assurance

Arcadia Supermarket
Tunbridge Wells

QUALITY PLAN

Document No: ASTW - QP - 01

Date: June 1988

6. INSPECTION, TESTING AND AUDITS (contd)

6.2 Records

Records of inspections and tests will be kept using the following forms:-

Excavation & fill:	Project Engineer to produce record form.
Concrete:	Readymix Order and Mix Details AC-164. Pre-placement Check and Inspection AC-263. Concrete Placement Record AC-264. Concrete Curing Record AC-265. Cube test results AC - 189.
Piling:	Use Sub-contractor's standard forms.
Brickwork:	Foreman to produce record form.
Steelwork:	To be defined in Sub-contractor's Work Instruction.

6.3 Audits

Quality Engineer will audit implementation of this Quality Plan at weeks 10, 20 and 30. Audit reports will be distributed as follows:

Project Manager
Company Quality Assurance Manager
Construction Manager

7. AMENDMENTS TO QUALITY PLAN

Review of this Quality Plan will be an agenda item at monthly site meetings. It will be re-issued at intervals of one month during the first six months of the project and thereafter when significant changes occur.

8. OTHER MEASURES

Project Manager to specify schedule of records to be maintained for long- and short-term retention.
Target date for issue: Week 4.

BIBLIOGRAPHY

Juran, J.M. (1964) *Management Breakthrough*, McGraw-Hill, Maidenhead.

Sayle, A.J. (1985) *Management Audits*, Allan J. Sayle.

Robson, M. (1986) *The Journey to Excellence*, John Wiley & Sons, Chichester.

Matfield, R. (ed.) (1984) *Quality Assurance in Nuclear Power Plants*, Harwood Academic Publishers, London.

Sasaki, N. and Hutchins, D. (eds) (1984) *The Japanese Approach to Product Quality*, Pergamon, Oxford.

INDEX